Lecture Notes in Physics

Volume 855

For further volumes:
www.springer.com/series/5304

The Lecture Notes in Physics

The series Lecture Notes in Physics (LNP), founded in 1969, reports new developments in physics research and teaching—quickly and informally, but with a high quality and the explicit aim to summarize and communicate current knowledge in an accessible way. Books published in this series are conceived as bridging material between advanced graduate textbooks and the forefront of research and to serve three purposes:

- to be a compact and modern up-to-date source of reference on a well-defined topic
- to serve as an accessible introduction to the field to postgraduate students and nonspecialist researchers from related areas
- to be a source of advanced teaching material for specialized seminars, courses and schools

Both monographs and multi-author volumes will be considered for publication. Edited volumes should, however, consist of a very limited number of contributions only. Proceedings will not be considered for LNP.

Volumes published in LNP are disseminated both in print and in electronic formats, the electronic archive being available at springerlink.com. The series content is indexed, abstracted and referenced by many abstracting and information services, bibliographic networks, subscription agencies, library networks, and consortia.

Proposals should be sent to a member of the Editorial Board, or directly to the managing editor at Springer:

Christian Caron
Springer Heidelberg
Physics Editorial Department I
Tiergartenstrasse 17
69121 Heidelberg/Germany
christian.caron@springer.com

Emilio Elizalde

Ten Physical Applications of Spectral Zeta Functions

Second Edition

 Springer

Prof. Dr. Emilio Elizalde
Institute of Space Science
Higher Council for Scientific Research
Bellaterra (Barcelona), Spain

ISSN 0075-8450   ISSN 1616-6361 (electronic)
Lecture Notes in Physics
ISBN 978-3-642-29404-4   ISBN 978-3-642-29405-1 (eBook)
DOI 10.1007/978-3-642-29405-1
Springer Heidelberg New York Dordrecht London

Library of Congress Control Number: 2012940223

Printed on acid-free paper

Springer is part of Springer Science+Business Media (www.springer.com)

To my dearest family,
for so many missed hours,
. . . and yet for a second time

Preface

The years elapsed since the First Edition was issued have witnessed an impressive evolution of some of the applications here considered, maybe not so of the mathematical background which constitutes the first part, which evolves at a much more steady pace. In writing this Second Edition the same guiding idea has been strictly respected to include as many applications as possible, at the price of explaining things in the most concise way still compatible with the comprehensibility of the different subjects treated.

The new text includes some additions to Chap. 1, specifically two sections on the general definition of the zeta function of a pseudo-differential operator and on operator regularization. In Chap. 4 a new Section has been included on extended Chowla–Selberg series formulas, associated with arbitrary forms of the type of a quadratic plus a linear plus a constant term. The second Section in Chap. 5, on the experimental verification of the Casimir effect, has been suppressed, for the account there had been rendered clearly obsolete by the many and very important advances in this field during the last fifteen years. A very brief, one-paragraph description of the last has been included at the end of the first Section. It is basically a guide to a number of relevant references where the reader will find the new developments in this presently very hot and rapidly growing field. In Chap. 6 a new application has been added, Sect. 6.5, on the treatment of scalar and vector fields on a spacetime with a noncommutative toroidal part. The title of the Chapter has been changed accordingly.

In Chap. 7 there is now a new Section on the combination of zeta and Hadamard regularizations, in relation with computations of the Casimir effect under realistic physical conditions. In Chap. 9 only the title has changed, to be consistent with the fact that an additional application was added to Chap. 6. Finally, Chap. 10 is new, and contains an application of the zeta function techniques to the regularization of the expected imprint of quantum vacuum fluctuations in cosmology, in particular, in relation with the cosmological constant problem. In all, the number of applications in this Second Edition has now raised to twelve, what could justify a change of the title of the book itself. However, being it just a second edition this would not be advisable, and has not been done.

Barcelona, Spain   E. Elizalde

Preface to the First Edition

This monography is, in the first place, a commented guide that invites the reader to plunge into the thrilling world of zeta functions and their applications in physics. Different aspects of this field of knowledge are considered, as one can see specifically in the table of contents.

The level of the book is elementary. It is intended for people with no or little knowledge of the subject. Everything is explained in full detail, in particular, the mathematical difficulties and tricky points, which too often constitute an insurmountable barrier for those who would have liked to become acquainted with that matter but never dared to ask (or did not manage to understand more complete, higher-level treatises). In this sense the present work is to be considered as a basic introduction to other books that have appeared recently.

Concerning the physical applications of the method of zeta-function regularization here described, quite a big choice is presented. The reader must be warned, however, that I have not tried to explain the underlying physical theories in complete detail (since this is undoubtedly out of scope), but rather to illustrate—simply and clearly—the precise way how the method must be applied. Sometimes zeta regularization is explicitly compared in the text with other procedures the reader is supposed to be more familiar with (as cut-off or dimensional regularization). Again, a very detailed comparison would have taken us too far away from the general purpose and, knowing the other procedures, the reader is already entitled to confront them directly. In the examples only physical systems with a known spectrum have been considered. This is the most simple case—although the procedure itself can be (and has been) extended to much more general situations.

I would like to thank my colleagues and friends Alfred Actor, Michael Bordag, Iver Brevik, Andrei Bytsenko, Guido Cognola, Klaus Kirsten, Yuri Kubyshin, Sergi Leseduarte, Sergei Odintsov, Sergi Rafels-Hildebrandt, August Romeo, Luciano Vanzo, and Sergio Zerbini. My gratitude goes to all of them for really enjoyable discussions and, in some cases, for specific contributions to the original papers on which part of this report is based.

Barcelona, Spain   E. Elizalde

Contents

1 Introduction and Outlook 1
   1.1 Zeta Functions 1
       1.1.1 The Riemann Zeta Function 1
       1.1.2 The Hurwitz Zeta Function 3
       1.1.3 The Epstein Zeta Function 4
       1.1.4 A Word on Related Bibliography 5
   1.2 Zeta Function Regularization 5
       1.2.1 Historical Background 5
       1.2.2 The Zeta Function of a Differential Operator 6
       1.2.3 Regularization of the Vacuum Energy 8
       1.2.4 Regularization of One-Loop Graphs 8
   1.3 Examples and a Comparison with Other Procedures 10
       1.3.1 Some Explicit Examples 10
       1.3.2 Comparison with Other Regularization Methods 11
       1.3.3 A Word of Warning 13
   1.4 Present Developments and a Point on Rigor 14
       1.4.1 Calculation of Heat-Kernel Coefficients 15
       1.4.2 Determinant of the Laplacian 16
   1.5 General Definition of the Zeta Function of a Pseudo-differential
        Operator 17
       1.5.1 The Zeta Function of a ΨDO 17
       1.5.2 The Zeta Regularized Determinant 18
       1.5.3 Multiplicative Anomaly 19
       1.5.4 On the Explicit Calculation of ζ_A and $\det_\zeta A$. 20
   1.6 Future Perspectives: Operator Regularization 21
       1.6.1 Generalization and Further Extensions 21

2 Mathematical Formulas Involving the Different Zeta Functions ... 23
   2.1 A Simple Recurrence for the Higher Derivatives of the Hurwitz
        Zeta Function 23
   2.2 The Zeta-Function Regularization Theorem 29

	2.2.1	The Theorem (Special Form)	31
2.3		Immediate Application of the Theorem	38
2.4		Expressions for Multi-series on Combinations Involving Arbitrary Constants and Exponents	41

3 A Treatment of the Non-polynomial Contributions: Application to Calculate Partition Functions of Strings and Membranes 49
3.1		Dealing with the Non-polynomial Term Δ_{ER}	49
	3.1.1	Proof of Asymptoticity of the Series	51
	3.1.2	The Remainder Term and the Poisson Resummation Formula	55
3.2		Numerical Estimates of the Remainder	57
3.3		Application: Summation of the String Partition Function for Different Ranges of the Temperature	60

4 Analytical and Numerical Study of Inhomogeneous Epstein and Epstein–Hurwitz Zeta Functions 67
4.1		Explicit Analytical Continuation of Inhomogeneous Epstein Zeta Functions	68
	4.1.1	The Particular Case of the Basic One-Dimensional Epstein–Hurwitz Series	69
	4.1.2	The Homogeneous Case: Chowla–Selberg's Formula	72
	4.1.3	Derivation of the General Two-Dimensional Formula	73
4.2		Extended Chowla–Selberg Formulas, Associated with Arbitrary Forms of Quadratic+Linear+Constant Type	77
	4.2.1	Limit $q \to 0$	78
	4.2.2	Case with $q = 0$ but $c_1 \neq 0$	79
	4.2.3	Case with $c_1 = \cdots = c_p = q = 0$	80
4.3		Numerical Analysis of the Inhomogeneous Generalized Epstein–Hurwitz Zeta Function	83
	4.3.1	Asymptotic Expansions of the Function and Its Derivatives with Respect to the Variable and Parameters	84

5 Physical Application: The Casimir Effect 95
5.1		Essentials of the Casimir Effect	95
	5.1.1	The Original Casimir Effect	95
	5.1.2	Connection with the van der Waals Forces and the London Theory	96
	5.1.3	The Specific Contribution of Casimir and Polder: Retarded van der Waals Forces	96
	5.1.4	The Lifshitz Theory	98
5.2		The Casimir Effect in Quantum Field Theory	100
	5.2.1	The Local Formulation of the Casimir Effect	100
	5.2.2	The Mystery of the Casimir Effect	100
	5.2.3	The Concept of the Vacuum Energy	101
	5.2.4	The Explicit, Regularized Definition of the Casimir Energy	103

5.2.5 Definition of the Casimir Energy Density and Its Relation
with the Vacuum Energy 104
5.3 A Very Simple Computation of the Casimir Effect 105
5.3.1 The Casimir Effect for a Free Massless Scalar Field
in $\mathbb{S}^1 \times \mathbb{R}^d$ and in $\mathbb{T}^2 \times \mathbb{R}^2$ Spacetimes 107
5.3.2 The Case of a Massless Scalar Field Between p
Perpendicular Pairs of Parallel Walls with Dirichlet
Boundary Conditions 110
5.3.3 Massless Scalar Field with Periodic and Neumann
Boundary Conditions, and Electromagnetic Field 114

6 Five Physical Applications of the Inhomogeneous Generalized
Epstein–Hurwitz Zeta Functions 119
6.1 Application: The Casimir Energy over Riemann Surfaces 120
6.2 Application: Kaluza–Klein Model with Spherical Compactification 125
6.3 Critical Behavior of a Field Theory at Non-zero Temperature . . . 131
6.4 Application to Quantizing Through the Wheeler–De Witt Equation 134
6.4.1 Explicit Zeta-Function Calculation of the Essential
Determinant and Extrema of the Potential 135
6.4.2 An Alternative Treatment by Means of Eisenstein Series . . 138
6.5 Spectral Zeta Function for Both Scalar and Vector Fields
on a Spacetime with a Noncommutative Toroidal Part 141
6.5.1 Poles of the Zeta Function 141
6.5.2 Explicit Analytic Continuation of $\zeta_\alpha(s)$, $\alpha = 2, 3$,
in the Complex s-Plane 143

7 Miscellaneous Applications Combining Zeta with Other
Regularization Procedures 147
7.1 Relation Between the Generalized Pauli–Villars and the Covariant
Regularizations 147
7.2 The Casimir Energy Corresponding to a Piecewise Uniform String 152
7.2.1 The Zero Temperature Theory 154
7.2.2 Regularized Casimir Energy and Numerical Results 157
7.2.3 The Finite Temperature Theory 160
7.3 Zeta and Hadamard Regularizations 164
7.3.1 A Zeta-Function Approach 166
7.3.2 Case of Two-Point Dirichlet Boundary Conditions 168
7.3.3 How to Deal with the Infinities? 169
7.3.4 Hadamard Regularization of the Casimir Effect 171

8 Applications to Gravity, Strings and p-Branes 175
8.1 Application to Spontaneous Compactification in Two-Dimensional
Quantum Gravity 175
8.2 Application to the Study of the Stability of the Rigid Membrane . . 178
8.2.1 Calculation of the Potential 179
8.2.2 The Limit of Large Spacetime Dimensionality 180

        8.2.3 A Saddle Point Analysis 182
        8.2.4 Explicit Expressions for the Zeta-Function Regularization
               of the Effective Potential 183
        8.2.5 Discussion of the General Case 185

**9   Eleventh Application: Topological Symmetry Breaking
      in Self-Interacting Theories** 189
    9.1   General Considerations 189
    9.2   The One-Loop Effective Potential for the Self-Interacting Theory . 190
    9.3   The One-Loop Topological Mass 194
    9.4   Renormalization of the Theory 196
    9.5   Symmetry Breaking Mechanism for a Massless Scalar Field 198

10 Twelfth Application: Cosmology and the Quantum Vacuum 201
    10.1 On the Reality of the Vacuum Fluctuations 202
    10.2 On the Curvature and Topology of Space 203
        10.2.1 On the Curvature 204
        10.2.2 On the Topology 204
    10.3 Vacuum Energy Fluctuations and the Cosmological Constant . . . 205
    10.4 Simple Model with Large and Small Compactified Dimensions . . 206
        10.4.1 Regularization of the Vacuum Energy Density 207
        10.4.2 Numerical Results 208
    10.5 Braneworld Models 209
    10.6 Supergraviton Theories 211

References 215

Index 225

Chapter 1
Introduction and Outlook

In this introductory chapter, an overview of the method of zeta function regularization is presented. We start with some brief historical considerations and by introducing the specific zeta functions that will be used in the following chapters in a number of physical situations, as the Riemann, Hurwitz (or Riemann generalized), and Epstein zeta functions. We summarize the basic properties of the different zeta functions. We define the concept of zeta function associated with an elliptic partial differential operator, and point towards its uses to define 'the determinant' of the operator, and we discuss the multiplicative anomaly or defect of the zeta determinant. We show explicitly how to regularize the Casimir energy in some simple cases in a correct way, thereby introducing the zeta-function regularization procedure. Finally, these fundamental concepts are both extended and made much more precise in the last section, where examples of the most recent developments on powerful applications of the theory are discussed. Further perspectives of this regularization method, as the so-called operator regularization, are provided at the very end.

1.1 Zeta Functions

1.1.1 The Riemann Zeta Function

The following identity (where s is real and p prime)

$$\prod_p \left(1 - p^{-s}\right)^{-1} = \sum_{n=1}^{\infty} \frac{1}{n^s}, \quad s > 1, \tag{1.1}$$

was first found by Leonhard Euler (1707–1783) while he was searching a proof of Euclid's theorem, that had led him to consider the divergence of the series of the reciprocals of all prime numbers. The last sum on the r.h.s. is a function of the real variable s (defined on the half-line $s > 1$) and the equality relates the behavior of the function with the properties of the prime numbers. In his famous work published in 1859, *Über die Anzahl der Primzahlen unter einer gegebenen Grösse*, Bernhard

E. Elizalde, *Ten Physical Applications of Spectral Zeta Functions*,
Lecture Notes in Physics 855,
DOI 10.1007/978-3-642-29405-1_1, © Springer-Verlag Berlin Heidelberg 2012

Riemann started from this identity, discovered by Euler more than one hundred years before, and realized that in order to be able to study the distribution of the primes he had to allow s to be a complex variable. He denoted this function of a complex variable by $\zeta(s)$, and it became known as 'the Riemann zeta function'[1]

$$\zeta(s) = \sum_{n=1}^{\infty} n^{-s}, \quad s \in \mathbb{C}, \ \operatorname{Re} s > 1. \tag{1.2}$$

Nowadays the Riemann zeta function is just one (albeit the most distinguished) member of a whole family of 'zeta functions' (Hurwitz, Epstein, Selberg, ...). Further, there exists the much more general concept of 'zeta function associated with a differential operator', as we shall see below (Sect. 1.4).

The series (1.2) is absolutely convergent in the open domain $\operatorname{Re} s > 1$ of the complex plane. Moreover, the integral representation

$$\zeta(s) = \frac{1}{\Gamma(s)} \int_0^{\infty} dt \, \frac{t^{s-1}}{e^t - 1} \tag{1.3}$$

shows that $\zeta(s)$ can be analytically continued everywhere in the complex s-plane except to the point $s = 1$, where it develops a pole of residue 1. In particular, one obtains that

$$\zeta(0) = -\frac{1}{2}, \quad \zeta(-2n) = 0, \quad \zeta(1 - 2n) = -\frac{B_{2n}}{2n}, \quad n = 1, 2, 3, \dots . \tag{1.4}$$

The Laurent series of $\zeta(s)$ around $s = 1$ is

$$\zeta(s) = \frac{1}{s-1} + \gamma + \gamma_1(s-1) + \gamma_2(s-1)^2 + \cdots,$$

$$\gamma_k = \lim_{n \to \infty} \left[\sum_{\nu=1}^{\infty} \frac{(\log \nu)^k}{\nu} - \frac{1}{k+1} (\log n)^{k+1} \right], \tag{1.5}$$

γ being Euler's constant.

An important property (common to all zeta functions) is the existence of a functional equation called also by physicists the *reflection formula*. For the Riemann zeta function it reads (as follows immediately from (1.3))

$$\zeta(s) = 2^s \pi^{s-1} \sin \frac{\pi s}{2} \Gamma(1-s) \zeta(1-s), \tag{1.6}$$

or, alternatively,

$$\Gamma\left(\frac{s}{2}\right) \zeta(s) = \pi^{s-1/2} \Gamma\left(\frac{1-s}{2}\right) \zeta(1-s). \tag{1.7}$$

---

[1] A hint for Spanish speaking colleagues. In Spanish (and some other languages) the Greek letter ζ is phonetically transcribed as *dseta* (to mimic its original Greek pronunciation). In Greek the letter which is actually pronounced as the Spanish *zeta* is θ.

As a result

$$\zeta(2n) = (-1)^{n+1}\frac{(2\pi)^{2n}}{2(2n)!}B_{2n}, \quad n = 1, 2, 3, \ldots, \tag{1.8}$$

where B_n are the Bernoulli numbers.

The only real zeros of the Riemann ζ are located at $s = -2, -4, -6, \ldots$. It is known that $\zeta(s)$ has no other zeros out of the strip $0 < \operatorname{Re} s < 1$ and that there are infinitely many complex zeros inside of it. The *Riemann conjecture* states that all the complex zeros of $\zeta(s)$ have real part equal to $\frac{1}{2}$. To obtain a proof of this conjecture was considered by Hilbert to be the most important open problem in mathematics [1]. Undoubtedly, this is a very important question related with the Riemann zeta function, for its consequences in number theory and other fields of knowledge, as information theory, chaotic systems, etc., and the number of new results originated in trying to answer it increase steadily. However, the applications to physical problems we are going to develop in this book have little to do with this issue.

1.1.2 The Hurwitz Zeta Function

An important generalization of ζ is the *Hurwitz zeta function* ζ_H (also called simply *generalized Riemann zeta function*)

$$\zeta_H(s, a) = \sum_{n=0}^{\infty}(n + a)^{-s}, \quad \operatorname{Re} s > 1, \ a \neq 0, -1, -2, \ldots. \tag{1.9}$$

An integral representation is

$$\zeta_H(s, a) = \frac{1}{\Gamma(s)}\int_0^{\infty} dt\, t^{s-1}\frac{e^{-ta}}{1 - e^{-t}}, \quad \operatorname{Re} s > 1, \ \operatorname{Re} a > 0. \tag{1.10}$$

It can be shown that $\zeta_H(s, a)$ has only one singularity—namely a simple pole at $s = 1$ with residue 1—and that it can be analytically continued to the rest of the complex s-plane. Its Laurent expansion around the pole is

$$\zeta_H(1 + \varepsilon, a) = \frac{1}{\varepsilon} - \psi(a) + O(\varepsilon), \tag{1.11}$$

with ψ the digamma function, and some special values are

$$\zeta_H(0, a) = \frac{1}{2} - a,$$

$$\zeta_H(-m, a) = -\frac{B_{m+1}(a)}{m + 1}, \quad m \in \mathbb{N}, \tag{1.12}$$

$B_r(a)$ being the Bernoulli polynomials.

The analogue of the Riemann zeta function for an unspecified algebraic number field, in number theory, is Dedekind's zeta function. The Riemann zeta function is actually the Dedekind's zeta function for the field of the rational numbers.

1.1.3 The Epstein Zeta Function

Another different extension of the Riemann zeta function is the whole family of Epstein zeta functions [2, 3]. Given a positive integer p, consider the vectors

$$
\begin{aligned}
\vec{g} &\equiv (g_1, \ldots, g_p), & g_i &\in \mathbb{R}, \\
\vec{h} &\equiv (h_1, \ldots, h_p), & h_i &\in \mathbb{R}, \\
\vec{m} &\equiv (m_1, \ldots, m_p), & m_i &\in \mathbb{R}.
\end{aligned}
\tag{1.13}
$$

Let $c = (c_{\mu\nu})$ be a non-singular symmetric $p \times p$ matrix, and φ its associated quadratic form

$$
\varphi(x) = \sum_{\mu,\nu=1}^{p} c_{\mu\nu} x_\mu x_\nu.
\tag{1.14}
$$

If the real part of $\varphi(x)$ is positive definite (for a complex variable s), the *Epstein zeta function* Z of order p, with characteristic $\left|\begin{matrix}\vec{g}\\\vec{h}\end{matrix}\right|$ and module $(c_{\mu\nu})$ is defined as [2, 4]

$$
Z\left|\begin{matrix}\vec{g}\\\vec{h}\end{matrix}\right|(s)_\varphi = Z\left|\begin{matrix}g_1\cdots g_p\\h_1\cdots h_p\end{matrix}\right|(s)_\varphi = \sum_{m_1,\ldots,m_p=-\infty}^{\infty}{}' \left[\varphi(\vec{m}+\vec{g})\right]^{-\frac{s}{2}} e^{2\pi i(\vec{m},\vec{h})}, \tag{1.15}
$$

where the prime means that when the $g_1, \ldots, g_p$ are integers, the values of $m_1, \ldots, m_p$ such that $\vec{m} + \vec{g} = \vec{0}$ have to be excluded from the summation. This series is absolutely convergent for $\operatorname{Re} s > p$, and defines an analytic function of s. The functional equation

$$
\pi^{-\frac{s}{2}}\Gamma\left(\frac{s}{2}\right) Z\left|\begin{matrix}\vec{g}\\\vec{h}\end{matrix}\right|(s)_\varphi = \Delta^{-\frac{1}{2}} e^{-2\pi i(\vec{g},\vec{h})} \pi^{-\frac{p-s}{2}}\Gamma\left(\frac{p-s}{2}\right) Z\left|\begin{matrix}\vec{h}\\-\vec{g}\end{matrix}\right|(p-s)_{\varphi^*} \tag{1.16}
$$

constitutes the keystone on which the theory of zeta functions rests, and yields all the special reflection formulas above as particular cases.

The function of s given in (1.15) is entire except when all components of $\vec{h}$ are integers. In that case the zeta function has a single pole at $s = p$, whose residue can be read from

$$
Z\left|\begin{matrix}\vec{g}\\\vec{h}\end{matrix}\right|(s)_\varphi = \frac{2\pi^{\frac{p}{2}}}{\Delta^{\frac{1}{2}}\Gamma(\frac{p}{2})}\frac{1}{s-p} + c_0 + c_1(s-p) + O\left((s-p)^2\right). \tag{1.17}
$$

The Epstein zeta function vanishes at even negative integers

$$Z \left| \begin{matrix} \vec{g} \\ \vec{h} \end{matrix} \right| (-2k)_\varphi = 0, \quad k = 1, 2, 3, \ldots \tag{1.18}$$

and also at $s = 0$, unless all the components of $\vec{g}$ are integers, in which case

$$Z \left| \begin{matrix} \vec{g} \\ \vec{h} \end{matrix} \right| (0)_\varphi = -e^{-2\pi i (\vec{g}, \vec{h})}. \tag{1.19}$$

Explicit values for $p = 1, 2$ (with $\vec{g} = \vec{h} = \vec{0}$) have been obtained in Epstein's papers. When $p = 1$ one recovers all the properties of the Riemann zeta function.

1.1.4 A Word on Related Bibliography

In Ref. [5] there is a nice summary of the properties that all the zeta functions have in common, together with a more extensive description of all kinds of zeta functions than the one given here. To mention just a few additional references on the subject let us recall the classical book by Titchmarsh [6], the ones by Edwards [7] and Ivić [8], by Jorgenson and Lang [9] and Elizalde et al. [10, 11], and the more recent ones by Karatsuba and Voronin [12], Kirsten [13], Bytsenko et al. [14], and Apostol [15].

1.2 Zeta Function Regularization

1.2.1 Historical Background

Possibly the first systematic contributions to the use of the zeta function to give sense to infinite series (what is called by physicists *regularization* and *renormalization*) are due to Godfrey H. Hardy and John E. Littlewood [16–18], starting from the second decade of last century. They actually established the convergence and equivalence of series regularized with the heat kernel and zeta function methods. As Hardy realized to his surprise, Srinivasa I. Ramanujan had also found for himself the functional equation of the zeta function. In the thirties, Torsten Carleman [19] went one step further, by constructing the zeta function encoding the eigenvalues of the Laplacian of a compact Riemannian manifold, for the case of a compact region of the plane.

A significant improvement, well recognized in the specialized literature, was achieved in 1949 by Subbaramiah Minakshisundaram and Åke Pleijel [20] who extended Carleman's results, showing that for the Laplacian of a compact Riemannian manifold, the corresponding zeta function converges and has an analytic continuation as a meromorphic function to the whole complex plane. In the middle sixties, Robert T. Seeley [21] extended these important results to elliptic pseudo-differential operators on compact Riemannian manifolds, showing that for such operators one can define the determinant using zeta function regularization. A few years later, Daniel B. Ray and Isadore M. Singer [22] used Seeley's results to define the determinant of a positive self-adjoint operator A (the Laplacian of a Riemannian manifold in their application), with eigenvalues $a_1, a_2, \ldots$, being in this case the zeta function formally the trace $\zeta_A(s) = \text{Tr}(A)^{-s}$, the same method defining the possibly divergent infinite product $\prod_{n=1}^{\infty} a_n = \exp[-\zeta_A'(0)]$.

After recognizing that all the above were very important steps in the quest towards the final formulation of a fully fledged zeta regularization method, it is now widely recognized by specialists that the honor to be the first to have clearly formulated such attempt where J. Stuart Dowker and Raymond Critchley in 1976 and Stephen Hawking in 1977. In their seminal work *Effective Lagrangian and energy-momentum tensor in de Sitter space* [23], Dowker and Critchley went definitely further in the application of the above procedures to physics: they actually proposed, for the first time, a zeta function regularization method for quantum physical systems. This paper has got high recognition, having gathered over 600 citations to present date. One thing specialists also point out is that, in spite of the fact that, elaborating from the methods developed in this paper, it is true that a well defined and clear regularization prescription for a general case can be easily obtained, these authors actually described the method very briefly in this work, the uses and wide possibilities of the procedure not having been fully exploited there. This is maybe the main reason why Hawking's extremely influential paper (it has got over 1100 citations up to date) entitled *Zeta function regularization of path integrals in curved spacetime* [24] is considered by many to be the actual seminal reference where the zeta function regularization method was defined, with all its computational power and possible physical applications, which were very clearly identified there. Needless to say, the title of the paper is absolutely explicit. After investigating the case in some detail, it is fair to conclude that the priority of Dowker and Critchley in this matter is now sufficiently well established in the literature.[2]

1.2.2 The Zeta Function of a Differential Operator

The method of zeta function regularization is schematically defined as follows. Take the Hamiltonian, H, corresponding to our quantum system, plus boundary conditions, plus possible background field and including a possibly non-trivial metric (because we may live in a curved spacetime). In mathematical terms, all this boils down to a (generically second order, elliptic, pseudo-) differential operator, A, plus

---

[2]With some incredible exceptions, however, as the current Wikipedia article on *"Zeta function regularization"*, where no mention to Dowker and Critchley is done!

corresponding boundary conditions. The spectrum of this operator A may or may not be calculable explicitly and, in the first case, may or may not exhibit a beautiful regularity in terms of powers of natural numbers. Under quite general conditions,[3] a *zeta function*, ζ_A, corresponding to the operator A, can be (almost uniquely) defined in a rigorous way. The *formal* expression of this definition is:

$$\zeta_A(s) = \operatorname{Tr} e^{-s \ln A}, \tag{1.20}$$

which will make sense just in some domain of the complex plane s and only if A fulfills typical spectral conditions. Let us stress that this definition is quite general in the sense that includes many interesting physical situations (see, however, below).

The zeta function $\zeta_A(s)$ is generically a meromorphic function (develops only poles) on the complex plane, $s \in \mathbb{C}$ [5, 25]. Its calculation usually requires complex integration around some circuit in the complex plane, the use of Mellin transforms, etc. and has very much to do (Sect. 1.4) with the calculation of invariants of the spacetime metric (the Hadamard–Minakshisundaram–Pleijel–Seeley–De Witt–Gilkey–... coefficients) [26–33].

In the particular case when the eigenvalues of the (pseudo-)differential operator A—or, what is equivalent, the eigenvalues of the Hamiltonian (with the boundary conditions taken into account)—can be calculated explicitly (let us call them λ_n and assume they form a discrete set, with n in general a multi-index, possibly with a continuous part), the expression of the zeta function is given by:

$$\zeta_A(s) = \sum_n \lambda_n^{-s}, \tag{1.21}$$

an expression valid to the rhs of the abscissa of convergence of the series (which value equals the dimension of the manifold divided by the order of the operator). This is then continued analytically to the rest of the complex plane, resulting in a meromorphic function. A very important point is that, in many cases, this analytical continuation is immediately provided by (a clever, suitable form of) the corresponding reflection formula (or functional equation, as is usually called by mathematicians) of the zeta function.[4] Notice that the generalization to the case of a continuous spectrum is quite simple (the multi-series being just substituted by a multiple integral). Now, as a particular case of this (already particular) case, when the eigenvalues are of one of the forms: (i) an, (ii) $a(n + b)$ or (iii) $a(n_1^2 + n_2^2)$, we obtain, respectively, the (i) (ordinary) Riemann zeta function ζ_R (or simply ζ), (ii) the Hurwitz (or generalized Riemann) zeta function ζ_H, and (iii) the Epstein zeta function Z (notice that in this last case the index n is double, or p-multiple in general, as we have seen before).

---

[3]We will give more precise specifications at the end of this chapter.

[4]An important portion of this book will be devoted to obtain such convenient forms of this reflection formula in different situations of physical interest.

1.2.3 Regularization of the Vacuum Energy

Depending on the physical magnitude to be calculated, the zeta function must be evaluated at a certain particular value of s. For instance, if we are interested in the vacuum or Casimir energy, which is simply obtained as the sum over the spectrum (for details see Chap. 5, Sect. 5.3)

$$E_C = \frac{\hbar}{2} \sum_n \lambda_n, \tag{1.22}$$

this will be given by the corresponding zeta function evaluated at $s = -1$:

$$E_C = \frac{\hbar}{2} \zeta_A(-1). \tag{1.23}$$

Normally, the series (1.22) will be divergent, and this formula involves an analytic continuation through the zeta function. That is why such regularization can be termed as a particular case of analytic continuation procedures. To understand this point better, let us just talk about the Riemann zeta function, $\zeta(s)$. As we have seen, it is given by the series expression $\zeta(s) = \sum_{n=1}^{\infty} n^{-s}$ in the region of the complex s-plane where $\mathrm{Re}\, s > 1$. In the rest of the complex plane $\zeta(s)$ is *not* given by the series (1.22) (which is *divergent*). But there exists one and only one meromorphic function that coincides with our initial series in the domain of convergence, and thus extends its definition to the rest of the plane—except for the point $s = 1$ on the real axis. This form of the domain of convergence, a half plane $\mathrm{Re}\, s > a$, for some $a \in R$, is typical of the so-called Dirichlet series and will appear also in the more complicated zeta functions we are going to consider in the next chapters. The value a is called the *abscissa of convergence* of the Dirichlet series. In general, when one talks about the zeta function of the operator A (usually the Hamiltonian, in physics) one certainly refers from the beginning to its analytical continuation to the whole complex plane, in the way just described.

1.2.4 Regularization of One-Loop Graphs

In perturbation theory, the determinant of an operator corresponds to a one-loop graph. One of the most widely used methods for extracting a finite value from that graph—in flat spacetimes—is dimensional regularization, in which the dimension d is treated like a complex variable, and whose analytic continuation up to $d = 4$ gives rise to a pole that will eventually be renormalized out. It was noted in [24] that it is not clear how to apply this procedure in curved spacetimes. As an example, if

one were handling $\mathbb{S}^4$—Euclidean version of the de Sitter space—the natural generalization would be $\mathbb{S}^d$ [34], but if one were to deal with the Schwarzschild solution,
with topology $\mathbb{R}^2 \times \mathbb{S}^2$, one might either extend it to $\mathbb{R}^2 \times \mathbb{S}^{d-2}$, or could add on extra dimensions to $\mathbb{R}^2$ as well. Then the value for a closed loop graph would depend
on the actual type of extension to d dimensions. It is in this sense that dimensional
regularization in curved spacetimes has been termed as *ambiguous*. In general, the
value given by zeta-function regularization coincides, up to a multiple of the nor-
malization parameter, with the one given by dimensional regularization when the
extra dimensions are flat.

In the study of black holes or, in general, in problems involving curved boundaries, the use of the asymptotic expansion for the operator heat kernel allows for
relating the behavior of the partition function—or, say, the generating functional—
under scale changes of the background spacetime to integrals of quadratic expressions in the curvature tensor. The energy-momentum tensor can be obtained by
functional differentiation of the partition function with respect to the background
metric, and can therefore be expressed in terms of derivatives of the heat kernel. As has long been known, the trace of the energy momentum tensor is related to the behavior of the suitable generating functional under scale transformations.

These observations were made by Hawking, but generalized zeta functions had
already been used in [23] in a manner that we would nowadays call regularization
of one-loop graphs, and which essentially consists in attaching zeta functions not to
path integrals, but to the Feynman diagrams themselves. A modern formulation of
this method has been developed in [35]. When evaluating path integrals in curved
spacetimes, one has to deal with expressions like

$$Z[g, \phi] = \int \mathcal{D}g \mathcal{D}\phi e^{iI[g,\phi]}, \tag{1.24}$$

where $\mathcal{D}g$ is a measure on the space of metrics, $\mathcal{D}\phi$ a measure on the space of
matter fields, and $I[g, \phi]$ the classical action. The integral is taken over all fields
g and ϕ satisfying certain boundary or periodicity conditions. For a canonical en-
semble at temperature $T = 1/\beta$ the boson fields are periodic in imaginary time on
some boundary at large distance, with period β. Z becomes the partition function
of statistical mechanics. The quadratic contribution in the field fluctuations reduces
to the evaluation of a determinant. However, as the convergence of the infinite product is by no means guaranteed, making sense of such expression is usually a non-
trivial matter. The free energy is proportional to the logarithm of Z, and has the
form

$$\log Z[\tilde{\phi}] = \frac{1}{2} \zeta'_{A_2}(0) + \frac{1}{2} \log\left(\frac{1}{2\pi\mu^2}\right) \zeta_{A_2}(0), \tag{1.25}$$

in terms of the zeta function of the quadratic part of the action and of its derivative
(for more details, see [10, 11, 14]).

1.3 Examples and a Comparison with Other Procedures

1.3.1 Some Explicit Examples

In Ref. [36], when calculating the Casimir energy of a piecewise uniform, closed string, the following expression arose [36, Eq. (52)]

$$\sum_{n=0}^{\infty} (n+\beta), \tag{1.26}$$

which is clearly infinite. Here, the zeta-function regularization procedure consists in the following. This expression comes about as the sum over the eigenvalues $n + \beta$ of the Hamiltonian of a certain quantum system (transverse oscillations of the mentioned string), i.e. $\lambda_n = n + \beta$. There is little doubt about what to do: as stated above, the corresponding zeta function is

$$\zeta_A(s) = \sum_{n=0}^{\infty} (n+\beta)^{-s}. \tag{1.27}$$

Now, for $\mathrm{Re}\, s > 1$ this is the expression of the Hurwitz zeta function $\zeta_H(s, \beta)$, which can be analytically continued as a meromorphic function to the whole complex plane. Thus, the zeta function regularization method unambiguously prescribes that the sum under consideration must be assigned the following value [37]

$$\sum_{n=0}^{\infty} (n+\beta) = \zeta_H(-1, \beta). \tag{1.28}$$

The wrong alternative (for obvious reasons after what has been said before), would be to argue that we might as well have written

$$\sum_{n=0}^{\infty} (n+\beta) = \zeta(-1) + \beta\zeta(0), \tag{1.29}$$

that gives a different result (a wrong one, in fact).

Of course, the method can be viewed as just one of the many possibilities of analytic continuation in some specific parameter in order to give sense to (i.e., to regularize) infinite expressions. From this point of view, it is related to the standard dimensional regularization method. Sometimes it has been argued that, being so close relatives, these two procedures even share the same type of diseases [38]. But precisely to cure an ambiguity problem, namely that of the dependence of the regularized result on the kind of the extra dimensions (artificially introduced in dimensional regularization) was one of the main motivations of Hawking for the introduction of a new procedure, i.e. zeta function regularization, in physics [23, 24, 39–41].

As is well known, a function $f(s)$ defined on a certain domain of the complex s-plane cannot have two different analytic continuations. However, one has to choose carefully the definition of this starting function $f(s)$ in the domain, namely the precise dependence of f on s. In fact, as exemplified in [38], if one starts from two different functions of s—which just coincide (formally) at the *only point* to which they are to be continued—the result one gets from the two functions is generally different. In particular,

$$f_1(s) = \sum_{n=0}^{\infty} n^{-s} \tag{1.30}$$

and

$$f_2(s) = \sum_{n=0}^{\infty} n \left(\frac{n}{a} + 1 \right)^{-(s+1)} \tag{1.31}$$

continued to $s = -1$, in the first example, which corresponds to a Hermitian massless conformal scalar field in 2d Minkowski spacetime with a compactified dimension, and

$$g_1(s) = \sum_{n=0}^{\infty} n^{-3s} \tag{1.32}$$

continued to $s = -1$, vs.

$$g_2(s) = \sum_{n=0}^{\infty} n^3 \left(\frac{n}{a} + 1 \right)^{-s}, \tag{1.33}$$

continued to $s = 0$, in the second example, in which the vacuum energy corresponding to a conformally coupled scalar field in an Einstein universe is studied. Needless to say, the number of different possibilities to proceed in this way is literally infinite. This shows the importance of adhering to the correct definition of zeta function associated with the differential operator corresponding to the physical observable one has to deal with—the Hamiltonian, in the case of the Casimir energy.

1.3.2 Comparison with Other Regularization Methods

Dimensional regularization is also an analytical continuation procedure. And, at first sight, to analytically continue in the number of dimensions of spacetime could seem an even more abstract and arbitrary method than continuation in a much less 'harmful' parameter as is a convenient exponent affecting the eigenvalues. Since, let us think for a moment, what physical meaning can be attributed to a number of dimensions $d = 3.5$ or $d = 4.1$? No doubt that one of the reasons for the success of dimensional regularization lies in the fact that in physics we are used to work in different number of dimensions: there are important theories in two, three and four dimensions, and having got an 'intuition' of what happens when one wanders from two to three and from three to four dimensions, it almost seems as if we could consider us capable of 'interpolating' the properties and ascertain with precision what is going to happen at $d = 3.5$ or $d = 4.1$. This may be a bit exaggerated, but the actual result is that we manage to feel quite 'at ease' with a non-integer (and even non-rational) number of dimensions. At the other extreme we can put cut-off regularization, undoubtedly the most 'physical' and, at the same time, mathematically the most arbitrary and less elegant of all regularization methods. It is quite easy to explain with simple arguments the physical meaning of a cut-off, in elementary terms understandable by any undergraduate student. For instance, the fact that, in the case of the standard Casimir effect, the plates are just transparent to the infinite number of high frequency modes, $\omega \geq \omega_{cut}$ (with ω_{cut} reflecting the scale of the plate separation), so that they do not contribute physically to the energy difference between the two situations of presence and absence of the plates, respectively. Often, it is not difficult to give a precise meaning to the invariance of our physical problem under the choice of a cut-off. In comparison, zeta-function regularization is mathematically much less simple than the choice of a cut-off and physically much less intuitive than cut-off or even dimensional regularization. However, once the technique of analytical continuation of the zeta series is mastered and confidence in the significance of the results of the method is secured (what is obtained by using the procedure and checking the results in a number of different situations), then the method of zeta-function regularization turns often to be invaluable.

Some points worth taking into account.

1. There could exist, in principle, infinitely many different analytic regularization procedures, being dimensional regularization and zeta function regularization just two of them.
2. Zeta function regularization is, to some extent, a uniquely defined procedure (one-loop, effective potentials, ...), through the concept of zeta function associated with the Hamiltonian. Sometimes, it even leads to a finite result, in which case it has been named *zeta function renormalization*. However, in general, a supplementary regularization followed by the corresponding renormalization procedure cannot be avoided (see Chap. 9).
3. Zeta function regularization does not suffer from some of the conceptual problems that afflict dimensional regularization (as the γ_5 problem or the question of the curvature of the 'additional' dimensions in a curved spacetime).
4. This does not mean, however, that zeta function regularization has *no* problems at all, but that they are, in fact, of a quite different kind. The first one appears already when it turns out that the point (let say $s = -1$ or $s = 0$) at which the zeta function must be evaluated turns out to be precisely a singularity (a *pole*) of the analytic continuation. This and similar difficulties can be solved, as discussed in detail in Ref. [42] (see also Chaps. 6 and 9 here). Eventually, as a final step one has to resort to renormalization group techniques (see [43] and Chap. 9 here).
5. Zeta function regularization has been extended to higher-loop order by McKeon and Sherry under the name of operator regularization [44, 45] (see also Shiekh

[46]). There, some serious difficulties, concerning the breakdown of gauge invariance appear [44, 45, 47] which, in practice, are possible to solve case by case and loop by loop (at least to two loops in a number of particular situations).

6. In the end, the fundamental question could be: which of the regularizations that are being used in physics is *the one* chosen by nature? (if this can be considered a valid question at all). In practice, one always tries to avoid answering it. The alternative response consists in just checking the finite results obtained with other more physically intuitive regularizations, or to compare them with some classical limits which provide well-known, physically meaningful values. We are obviously led to believe that in view of its simplicity and mathematical elegance, zeta function regularization could well be *the* regularization of the future. Those properties are certainly to be counted among their main virtues, but (oddly enough) in some sense also as its main intellectual drawbacks: we do not manage to see in detail which infinites are thrown away and when and how this happens, something that is handy in other more pedestrian regularizations— which are actually equivalent in most cases to the zeta one (as pointed out, e.g., in [38]).

1.3.3 A Word of Warning

Aside from some very simple cases (among those, the ones reviewed here), the use of the procedure of analytic continuation through the zeta function requires a good deal of heavy mathematical work [10, 11, 13, 14]. It is then no surprise that it has been so often associated with mistakes and sound errors (see [48] for some specific details), coming mainly from dubious manipulations. Some possible mistakes have been described before and some others will arise in the next chapters. Let us here just give an example of dubious manipulation. One which often appears starts from the well-known expression (see, for instance, [49])

$$\frac{a^2}{\pi^2} \sum_{n=1}^{\infty} \left(\pi^2 n^2 + \frac{a^2 m^2}{\pi^2} \right)^{-1} = \frac{1}{2m^2} \left(-1 + \frac{am}{\pi} \coth \frac{am}{\pi} \right), \qquad (1.34)$$

in other words (for $a = \pi$ and $m = c$),

$$\sum_{n=1}^{\infty} \left(\pi^2 n^2 + c^2 \right)^{-1} = \frac{1}{2c^2} (-1 + c \coth c). \qquad (1.35)$$

The integrated version of this equality, namely,

$$\sum_{n=-\infty}^{\infty} \ln \left(n^2 + \frac{c^2}{\pi^2} \right) = 2c + 2 \ln \left(1 - e^{-2c} \right), \qquad (1.36)$$

under the specific form

$$T \sum_{n=-\infty}^{\infty} \ln\left[(\omega_n)^2 + (q_l)^2\right] = q_l + 2T \ln\left(1 - e^{-q_l/T}\right), \tag{1.37}$$

with $\omega_n = 2\pi nT$ and $q_l = \pi l/R$, has been used in Ref. [50, Eq. (2.20)] when study-ing the Nambu–Goto string model at finite length and non-zero temperature. Now this equality is again formal. It involves an analytic continuation, since it has no sense to integrate the left hand side term by term: we get a divergent series.

A possible way to proceed is as follows. The expression on the left hand side hap-pens to be the most simple form of the inhomogeneous Epstein zeta function [51]. This function is quite involved and different expressions for it (including asymp-totical expansions very useful for accurate numerical calculations) will be given in Chap. 4 (see [51] and also [52]). In particular

$$\zeta_{EH}\left(s; c^2\right) = \sum_{n=1}^{\infty} \left(n^2 + c^2\right)^{-s}$$

$$= -\frac{c^{-2s}}{2} + \frac{\sqrt{\pi}\,\Gamma(s - 1/2)}{2\Gamma(s)} c^{-2s+1}$$

$$+ \frac{2\pi^s c^{-s+1/2}}{\Gamma(s)} \sum_{n=1}^{\infty} n^{s-1/2} K_{s-1/2}(2\pi nc), \tag{1.38}$$

which is reminiscent of the famous Chowla–Selberg formula [25]. Derivatives can be taken here and the analytical continuation in s presents again no problem (Chap. 4).

The usefulness of zeta function regularization is without question [5, 10, 11, 13, 14, 53, 54]. It can give immediate sense to expressions such as $1 + 1 + 1 + \cdots = -1/2$ (also obtainable by other means, of course), which turn out to be invaluable for the construction of new physical theories, as different as Pauli–Villars regular-ization with infinite constants (advocated by Slavnov [55], see Chap. 7) and matter generation in cosmology (see Chap. 8).

1.4 Present Developments and a Point on Rigor

We shall here describe, for completeness, some recent developments that have ex-tended the scope of applications of the zeta function method. They are of great potential importance but, in spite of this, in the applications to be described in the following chapters we are not going to make use of them since, as the title of the book clearly indicates, we shall restrict ourselves to the case when the spectrum of the differential operator is explicitly known and gives rise to some of the stan-dard zeta functions described before. Precise conditions for the definition of the zeta function of a differential operator will be given at the end of the section.

1.4.1 Calculation of Heat-Kernel Coefficients

An important issue for many years was to obtain explicitly the coefficients which appear in the short-time expansion of the heat-kernel $K(t)$ corresponding to a Laplacian-like operator of a d-dimensional manifold $\mathcal{M}$. In mathematics this interest originates, in particular, in the well-known connections that exist between the heat-equation and the Atiyah–Singer index theorem. In physics, the importance of that expansion is notorious in different domains of quantum field theory, where it is commonly known as the (integrated) Schwinger–De Witt proper-time expansion. In this context, the heat-equation for an elliptic (in general pseudo-elliptic) differential operator A and the corresponding zeta function $\zeta_A(s)$ have been realized to be particularly useful tools for the determination of effective actions [24] and for the calculation of vacuum or Casimir energies [42] (a basic issue for understanding the vacuum structure of a quantum field theory). Here usually the derivative $\zeta_A'(0)$ of the zeta function [24] and its value at $s = -1/2$ are needed.

A very quick and powerful method for the calculation of heat-kernel coefficients makes use of quite common ideas, as integral representations of the spectral sum, Mellin transforms, non-trivial commutation of series and integrals and skilful analytic continuation of zeta functions on the complex plane [56]. The method can be applied to the case of the heat-kernel expansion of the Laplace operator on a d-dimensional ball with either Dirichlet, Neumann or, in general, Robin boundary conditions. The final formulas obtained with this method are quite simple. The scheme has been illustrated in all detail in Ref. [56]. It serves for the calculation of an (in principle) arbitrary number of heat-kernel coefficients in any situation when the basis functions are known. In that paper, a complete list of new results has been given for the coefficients $B_3, \ldots, B_{10}$, corresponding to the d-dimensional ball with all the mentioned boundary conditions and $d = 3, 4, 5$.

In order to obtain these results, a specific property of the zeta function $\zeta_A(s)$ corresponding to an elliptic operator A is exploited, namely its well-known close connection with the heat-kernel expansion. In spite of the fact that almost everybody is aware of such connection, its actual use in the literature has remained very restricted till now. If the manifold $\mathcal{M}$ has a boundary $\partial\mathcal{M}$, the coefficients B_n in the short-time expansion have both a volume and a boundary part. It is usual to write this expansion in the form

$$K(t) \sim (4\pi t)^{-\frac{D}{2}} \sum_{k=0,1/2,1,\ldots}^{\infty} B_k t^k, \tag{1.39}$$

with

$$B_k = \int_{\mathcal{M}} dV \, b_n + \int_{\partial\mathcal{M}} dS \, c_n. \tag{1.40}$$

For the volume part very effective systematic schemes have been developed (see for example [57–59]). The calculation of c_n, however, is in general more difficult and this is precisely the strongest point of the new method based on the zeta-function calculations.

The connection between the heat-kernel expansion, (1.39) and the associated zeta function is established through the formulas [60]

$$\text{Res}\,\zeta(s) = \frac{B_{\frac{m}{2}-s}}{(4\pi)^{\frac{m}{2}}\Gamma(s)}, \tag{1.41}$$

for $s = \frac{m}{2}, \frac{m-1}{2}, \dots, \frac{1}{2}, -\frac{2l+1}{2}$, for $l \in \mathbb{N}_0$, and

$$\zeta(-p) = (-1)^p p! \frac{B_{\frac{m}{2}+p}}{(4\pi)^{\frac{m}{2}}}, \tag{1.42}$$

for $p \in \mathbb{N}_0$. It has been shown in [56] that these equations, (1.41) and (1.42), can actually serve as a very convenient starting point for the calculation of the coefficients B_k, even in the cases when the eigenvalues of the operator P under consideration are *not* known. The extensive knowledge in explicit zeta-function evaluations that has been accumulated in the past few years—which has its fundament in the formulas that will be presented and applied in the following chapters—has allowed to elaborate this competitive method of calculation of the heat-kernel coefficients.

Earlier investigations used Laplace transformations of the heat-kernel $K(t)$ itself, but there an intermediate cut off had to be introduced at some point—because one needed to consider the Laplace transform of a function which is singular at $t = 0$. In contrast, in the new approach it is the complex argument s of the zeta function of the Laplace operator which very neatly serves for the regularization of all sums (in just the usual way, see Sect. 1.2 above).

1.4.2 Determinant of the Laplacian

Another interesting development is the following. Motivated by the need to give answers to some basic questions in quantum field theory, during the last years there has been (and continues to be) a lot of interest in the problem of calculating the determinant of a differential operator, A (see for example [61, 62]). Often one has to deal in these situations with positive elliptic differential operators acting on sections of a vector bundle over a compact manifold. In such cases A has a discrete spectrum $\lambda_1 \leq \lambda_2 \leq \dots \to \infty$. The determinant, $\det A = \prod_i \lambda_i$, is generally divergent and one needs to make sense out of it by means of some kind of analytic continuation. A most appropriate way of doing that is by using the zeta function regularization prescription introduced by Ray and Singer [22] (see also [23, 24]). In this procedure $\ln \det A$ is defined by analytically continuing the function $\sum_i \lambda_i^{-s} \ln \lambda_i$ in the exponent s, from the domain of the complex plane where the real part of s is large to the point $s = 0$. Introducing the zeta function associated with the spectrum λ_i of A, $\zeta_A(s) = \sum_i \lambda_i^{-s}$, this is equivalent to defining

$$\ln \det A = -\zeta_A{}'(0).$$

Only a few general methods for the evaluation of $\ln \det A$ are available. Thus, for example, given that the manifold has a boundary, in [63] the determinants of differential and difference operators have been related to the boundary values of solutions of the operators. When A is a conformally covariant differential operator, exact results may sometimes be obtained by transforming to a 'more simple' operator $\tilde{A}$ for which $\ln \det \tilde{A}$ is known. Then, the knowledge of the associated heat-kernel coefficients nowadays available gives sometimes the exact value of $\ln \det A$ [64]. This approach has been used by Dowker to find the functional determinants for a variety of sectors of Euclidean space, spheres and flat balls for dimensions $D \leq 4$ [65–67]. Similar techniques have proven to be very powerful in order to obtain estimates of different types.

As a rule, however, explicit knowledge of the eigenvalues λ_i is necessary in order to obtain $\ln \det A$. This explicit knowledge of the eigenvalues is in general only guaranteed for highly symmetric regions of space, such as the torus, sphere or regions bounded by parallel planes. For these manifolds, detailed calculations have been performed in the context of Casimir energies and effective potential considerations. Those will be the cases to be investigated in the rest of this book.

In Ref. [64], the same method described before has been extended to the calculation of functional determinants of the Laplace operator on balls. Again, Dirichlet and Robin boundary conditions have been considered and, using this approach, formulas for any value of the dimension d of the ball have been obtained explicitly, for dimensions $d = 2, 3, 4, 5$ and 6. They can be easily extended to any value of d. That paper focuses on a class of situations for which the eigenvalues of the operator are not known explicitly but for which, nevertheless, the calculation of $\ln \det A$ is possible. The method developed is applicable whenever an implicit equation satisfied by the eigenvalues is known and some properties (later specified) of this equation are known too. The approach is exemplified by taking $A = -\Delta$ on the D-dimensional ball $B^d = \{x \in \mathbb{R}^d; |x| \leq r\}$, together with Dirichlet—or general Robin—boundary conditions.

1.5   General Definition of the Zeta Function of a Pseudo-differential Operator

1.5.1   The Zeta Function of a ΨDO

The *zeta function* ζ_A of A, a positive-definite elliptic pseudo-differential operator (ΨDO) of positive order $m \in \mathbb{R}$ (acting on the space of smooth sections of E, an n-dimensional vector bundle over a closed n-dimensional manifold, M) is defined as

$$\zeta_A(s) = \operatorname{tr} A^{-s} = \sum_j \lambda_j^{-s}, \quad \operatorname{Re} s > \frac{n}{m} \equiv s_0. \tag{1.43}$$

being $s_0 = \dim M / \operatorname{ord} A$ the *abscissa of convergence* of $\zeta_A(s)$. It can be proven that $\zeta_A(s)$ has a meromorphic continuation to the whole complex plane $\mathbb{C}$ (regular at $s = 0$), provided the principal symbol of A $(a_m(x, \xi))$ admits a *spectral cut*: $L_\theta = \{\lambda \in \mathbb{C}; \operatorname{Arg} \lambda = \theta, \theta_1 < \theta < \theta_2\}$, Spec $A \cap L_\theta = \emptyset$ (Agmon–Nirenberg condition [68]). This definition of $\zeta_A(s)$ depends on the position of the cut L_θ. The *only possible* singularities of $\zeta_A(s)$ are *simple poles* at $s_k = (n - k)/m, k = 0, 1, 2, \ldots, n - 1, n + 1, \ldots$. M. Kontsevich and S. Vishik have managed to extend this definition to the case when $m \in \mathbb{C}$ (no spectral cut exists) [68].

1.5.2 The Zeta Regularized Determinant

Let A be a ΨDO operator with a spectral decomposition: $\{\varphi_i, \lambda_i\}_{i \in I}$, with I some set of indices. The definition of determinant starts by trying to make sense of the product $\prod_{i \in I} \lambda_i$, which can be easily transformed into a "sum": $\ln \prod_{i \in I} \lambda_i = \sum_{i \in I} \ln \lambda_i$. From the definition of the zeta function of A: $\zeta_A(s) = \sum_{i \in I} \lambda_i^{-s}$, by taking the derivative at $s = 0$: $\zeta_A'(0) = -\sum_{i \in I} \ln \lambda_i$, we arrive to the following definition of determinant of A [22, 69, 70]:

$$\det{}_\zeta A = \exp\left[-\zeta_A'(0)\right]. \tag{1.44}$$

An older definition (due to Weierstrass) is obtained by subtracting in the series above (when it is such) the leading behavior of λ_i as a function of i, as $i \to \infty$, until the series $\sum_{i \in I} \ln \lambda_i$ is made to converge [71]. The shortcoming—for physical applications—is here that these additional terms turn out to be *non-local* and, thus, are non-admissible in a renormalization procedure.

In algebraic QFT, to write down an action in operator language one needs a functional that replaces integration. For the Yang–Mills theory this is the Dixmier trace, which is the *unique* extension of the usual trace to the ideal $\mathcal{L}^{(1,\infty)}$ of the compact operators T such that the partial sums of its spectrum diverge logarithmically as the number of terms in the sum: $\sigma_N(T) \equiv \sum_{j=0}^{N-1} \mu_j = \mathcal{O}(\log N), \mu_0 \geq \mu_1 \geq \cdots$. The definition of the Dixmier trace of T is: $\operatorname{Dtr} T = \lim_{N \to \infty} \frac{1}{\log N} \sigma_N(T)$, provided that the Cesaro means $M(\sigma)(N)$ of the sequence in N are convergent as $N \to \infty$ (remember that: $M(f)(\lambda) = \frac{1}{\ln \lambda} \int_1^\lambda f(u) \frac{du}{u}$). Then, the Hardy–Littlewood theorem can be stated in a way that connects the Dixmier trace with the residue of the zeta function of the operator T^{-1} at $s = 1$ (see Connes [72]): $\operatorname{Dtr} T = \lim_{s \to 1^+} (s - 1) \zeta_{T^{-1}}(s)$.

The Wodzicki (or noncommutative) residue [73] is the *only* extension of the Dixmier trace to the ΨDOs which are not in $\mathcal{L}^{(1,\infty)}$. It is the *only* trace one can define in the algebra of ΨDOs (up to a multiplicative constant), its definition being: $\operatorname{res} A = 2 \operatorname{Res}_{s=0} \operatorname{tr}(A \Delta^{-s})$, with Δ the Laplacian. It satisfies the trace condition: $\operatorname{res}(AB) = \operatorname{res}(BA)$. A very important property is that it can be expressed as an integral (local form) $\operatorname{res} A = \int_{S^*M} \operatorname{tr} a_{-n}(x, \xi) \, d\xi$ with $S^*M \subset T^*M$ the co-sphere bundle on M (some authors put a coefficient in front of the integral: Adler–Manin residue).

If dim $M = n = -\,\mathrm{ord}\,A$ (M compact Riemann, A elliptic, $n \in \mathbb{N}$) it coincides with the Dixmier trace, and one has $\mathrm{Res}_{s=1}\,\zeta_A(s) = \frac{1}{n}\,\mathrm{res}\,A^{-1}$. The Wodzicki residue continues to make sense for ΨDOs of arbitrary order and, even if the symbols $a_j(x,\xi)$, $j < m$, are not invariant under coordinate choice, their integral is, and defines a trace. All residua at poles of the zeta function of a ΨDO can be easily obtained from the Wodzicki residue [74].

1.5.3 Multiplicative Anomaly

Given A, B and AB ΨDOs, even if ζ_A, ζ_B and ζ_{AB} exist, it turns out that, in general, $\det_\zeta (AB) \neq \det_\zeta A \det_\zeta B$. The multiplicative (or noncommutative, or determinant) anomaly is defined as:

$$\delta(A, B) = \ln\left[\frac{\det_\zeta(AB)}{\det_\zeta A \det_\zeta B}\right] = -\zeta'_{AB}(0) + \zeta'_A(0) + \zeta'_B(0). \qquad (1.45)$$

Wodzicki's formula for the multiplicative anomaly [73, 75]:

$$\delta(A, B) = \frac{\mathrm{res}\{[\ln \sigma(A, B)]^2\}}{2\,\mathrm{ord}\,A\,\mathrm{ord}\,B(\mathrm{ord}\,A + \mathrm{ord}\,B)}, \qquad \sigma(A, B) := A^{\mathrm{ord}\,B}B^{-\mathrm{ord}\,A}. \quad (1.46)$$

At the level of Quantum Mechanics (QM), where it was originally introduced by Feynman, the path-integral approach is just an alternative formulation of the theory. In QFT it is much more than this, being in many occasions *the* actual formulation of QFT [76]. In short, consider the Gaussian functional integration

$$\int [d\Phi]\exp\left\{-\int d^D x\left[\Phi^\dagger(x)(\ )\Phi(x) + \cdots\right]\right\} \longrightarrow \det(\ )^{\pm 1}, \qquad (1.47)$$

(the sign $\pm$ depends on the spin-class of the integration fields) and assume that the operator matrix has the following simple structure (being each A_i an operator on its own):

$$\begin{pmatrix} A_1 & A_2 \\ A_3 & A_4 \end{pmatrix} \longrightarrow \begin{pmatrix} A & \\ & B^r \end{pmatrix}, \qquad (1.48)$$

where the last expression is the result of diagonalizing the operator matrix. A question now arises. What is the determinant of the operator matrix: $\det(AB)$ or $\det A \cdot \det B$? This has been very much on discussion during the last months [77–80]. There is agreement in that: (i) In a situation where a superselection rule exists, AB has no sense (much less its determinant), and then the answer must be $\det A \cdot \det B$. (ii) If the diagonal form is obtained after a change of basis (diagonalization process), then the quantity that is preserved by such transformations is the value of $\det(AB)$ and *not* the product of the individual determinants (there are counterexamples supporting this viewpoint [81–83]).

1.5.4 On the Explicit Calculation of ζ_A and $\det_\zeta A$

A fundamental property of many zeta functions is the existence of a reflection formula (or functional equation, in the mathematical language). For the Riemann zeta function: $\Gamma(s/2)\zeta(s) = \pi^{s-1/2}\Gamma(1 - s/2)\zeta(1 - s)$. For a generic zeta function, $Z(s)$, it is $Z(\omega - s) = F(\omega, s)Z(s)$, and allows for its analytic continuation in an easy way—what is, as advanced above, the whole story of the zeta function regularization procedure (at least the main part of it). But the analytically continued expression thus obtained is just another series, again with a slow convergence behavior, of power series type [84] (actually the same that the original series had, in its own domain of validity). S. Chowla and A. Selberg found a formula, for the Epstein zeta function in the two-dimensional case [85], that yields *exponentially quick convergence, and not only in the reflected domain*. They were extremely proud of that formula—as one can appreciate just reading the original paper (where actually no hint about its derivation was given, see [85]). In Ref. [86], I generalized this expression to inhomogeneous zeta functions (most important for physical applications), but staying always in *two* dimensions, for this was commonly believed to be an unsurmountable restriction of the original formula (see, e.g., Ref. [5]). I have obtained an extension to an *arbitrary* number of dimensions [87], both in the homogeneous (quadratic form) and non-homogeneous (quadratic plus affine form) cases.

In short, for the following zeta functions (corresponding to the general quadratic—plus affine—case and to the general affine case, in any number of dimensions, d) explicit formulas of the CS type were obtained in [87], namely,

$$\zeta_1(s) = \sum_{\vec{n}\in\mathbb{Z}^d} \left[Q(\vec{n}) + A(\vec{n})\right]^{-s} \qquad (1.49)$$

and

$$\zeta_2(s) = \sum_{\vec{n}\in\mathbb{N}^d} A(\vec{n})^{-s}, \qquad (1.50)$$

where Q is a non-negative quadratic form and A a general affine one, in d dimensions (giving rise to Epstein and Barnes zeta functions, respectively). Moreover, expressions for the more difficult cases when the summation ranges are interchanged, that is:

$$\zeta_3(s) = \sum_{\vec{n}\in\mathbb{N}^d} \left[Q(\vec{n}) + A(\vec{n})\right]^{-s} \qquad (1.51)$$

and

$$\zeta_4(s) = \sum_{\vec{n}\in\mathbb{Z}^d} A(\vec{n})^{-s} \qquad (1.52)$$

have been given in [87].

1.6 Future Perspectives: Operator Regularization

The Operator Regularization (OR) approach, due originally to D.G.C. McKeon and T.N. Sherry [88, 89] is considered as a genuine generalization of the zeta regularization approach. Its main aim is to extend zeta regularization, so effective at one-loop order [90], to higher loops. It has a distinct advantage over other competing procedures, in that it can be used with formally non-renormalizable theories, as shown in [91, 92]. A further feature of this approach is that divergences are not reabsorbed, each one is removed and replaced by an arbitrary factor. Indeed, operator regularization (OR) does not cure the non-predictability problem of non-renormalizability, but an advantage of the method is that the initial Lagrangian does not need to be extended with the addition of extra terms. The OR scheme is governed by the identity:

$$H^{-m} = \lim_{\epsilon \to 0} \frac{d^n}{d\epsilon^n} \left[1 + \left(1 + \alpha_1 \epsilon + \alpha_2 \epsilon^2 + \cdots + \alpha_n \epsilon^n \right) \frac{\epsilon^n}{n!} H^{-\epsilon-m} \right], \quad (1.53)$$

where the α_i's are arbitrary, and it is enough that the degree of regularization is equal to the loop order, n.

Two separate aspects of the procedure are, first the regularization itself and, second, the analytical continuation, where divergences are replaced by arbitrary factors. Thus, the effect of OR is in the end replace the divergent poles by arbitrary constants, as

$$\frac{1}{\epsilon^n} \longrightarrow \alpha_n, \quad (1.54)$$

to yield the finite expression

$$H^{-m} = \alpha_n c_{-n} + \cdots + \alpha_1 c_{-1} + c_0. \quad (1.55)$$

1.6.1 Generalization and Further Extensions

The OR method can be generalized to *multiple operators*, as in multi-loop cases

$$H^{-m_1} \cdots H^{-m_r} = \lim_{\epsilon \to 0} \frac{d^n}{d\epsilon^n} \left[1 + \left(1 + \alpha_1 \epsilon + \alpha_2 \epsilon^2 + \cdots + \alpha_n \epsilon^n \right) \right.$$
$$\left. \cdot \frac{\epsilon^n}{n!} H^{-\epsilon-m_1} \cdots H^{-\epsilon-m_r} \right]. \quad (1.56)$$

Further extensions of the procedure have been proposed. Let us recall that OR was first introduced in the context of the Schwinger approach,

$$\ln H = -\lim_{\epsilon \to 0} \frac{d^n}{d\epsilon^n} \left(\frac{\epsilon^{n-1}}{n!} H^{-\epsilon} \right), \quad (1.57)$$

which is known to be equivalent to the Feynman one

$$H^{-m} = \lim_{\epsilon \to 0} \frac{d^n}{d\epsilon^n} \left(\frac{\epsilon^n}{n!} H^{-\epsilon - m} \right). \tag{1.58}$$

The Schwinger form can be transformed into the Feynman one, as

$$H^{-m} = \frac{(-1)^{m-1}}{(m-1)!} \frac{d^m}{dH^m} \ln H. \tag{1.59}$$

Equivalence with dimensional regularization can be established in many cases, but not always. Problems, the main one being unitarity, may appear (see [93]). To start with, its naive application to obtain finite amplitudes breaks unitarity.

A definite advantage of the procedure is that, actually, no symmetry-breaking regulating parameter is ever inserted into the initial Lagrangian [94]. One can use Bogoliubov's recursion formula in order to show how to construct a consistent OR operator, and unitarity is upheld by employing a generalized evaluator consistently including lower-order quantum corrections to the quantities of interest. Unitarity requirements lead to *unique* expressions for quantum field theoretic quantities, order by order in $\hbar$. This fact has been proven in many cases (as for the Φ^4 theory at two-loop order, etc.). But I should say that, to my knowledge, a universal proof of this issue is actually still missing.

A final comment is in order. Using a BPHZ-like scheme, as the above one turns out to be, in the end, essentially reintroduces counterterms into the procedure, since they are actually hidden in the subtractions taking place at each step. In this way, the simplicity of the original zeta function regularization procedure, as described in the previous sections, and which is one of its main characteristics, is absent in the extended, operator regularization method.

Chapter 2
Mathematical Formulas Involving the Different Zeta Functions

In this chapter, a compendium of original formulas resulting from the zeta-regularization techniques, developed by the author and collaborators is given. Although some of the original derivations are reproduced, what follows is mainly intended as a table for practical use by the reader—the full derivations and arguments involved can be found in the accompanying bibliographical references. In particular, useful expressions are provided for the analytic continuation of Riemann, Hurwitz and Epstein zeta functions and generalizations of them, for their asymptotic expansions (including those for derivatives of Hurwitz's ζ), the zeta-function regularization theorem—and its use for multiple zeta-functions with arbitrary exponents—and, in another section, the first immediate applications of the theorem. All this is followed by a very careful study of the analytic continuation of multiple series which terms are combinations involving arbitrary coefficients and exponents, a case that is very involved and has not been treated properly in the mathematical literature. Of course this case always involves the elusive term that shows up in the correct application of the zeta-function regularization theorem. Some mistakes which regretfully appeared in a few formulas of the original papers have been corrected.

2.1 A Simple Recurrence for the Higher Derivatives of the Hurwitz Zeta Function

A recurrent formula which allows for the calculation of the asymptotic series expansion of any derivative, $\zeta^{(m)}(z,a) = \partial^m \zeta(z,a)/\partial z^m$, of the Hurwitz zeta function $\zeta(z,a)$ is here given. In particular, the first terms of the series corresponding to $\zeta''(-n,a)$ in inverse powers of a are written explicitly, for $n = 0, 1, 2, 3$. Knowledge of these expressions is basic in the zeta-function regularization procedure.

Some time ago, an asymptotic expansion for the first derivative

$$\zeta'(-n,a) \equiv \frac{\partial}{\partial z}\zeta(z,a)\bigg|_{z=-n}, \quad n = 0, 1, 2, \ldots, \tag{2.1}$$

E. Elizalde, *Ten Physical Applications of Spectral Zeta Functions*,
Lecture Notes in Physics 855,
DOI 10.1007/978-3-642-29405-1_2, © Springer-Verlag Berlin Heidelberg 2012

of the Hurwitz zeta function

$$\zeta(z,a) = \sum_{n=0}^{\infty}(n+a)^{-z}, \quad \mathrm{Re}\,z > 1, \; a \neq 0, -1, -2, \ldots \tag{2.2}$$

in inverse powers of a was derived [95] (see (2.8)), that has been found to be very useful by a number of authors—as a convenient tool, e.g., for the computation of effective actions in non-trivial backgrounds and also for the derivation of other interesting zeta function relations (see, for example, Steiner [96], Rudaz [97], and Ref. [51]). The simplicity and very quick convergence of the expressions we will give below have been recognized to be their most remarkable characteristics. They render them very useful for numerical applications [96], and for the subsequent derivation of related expressions for other zeta [97] and theta [51] functions.

In this chapter we describe the use the procedure of Refs. [95, 98] in order to obtain the asymptotic expansion corresponding to any derivative of the Hurwitz zeta function (2.2)

$$\zeta^{(m)}(z,a) \equiv \frac{\partial^m}{\partial z^m}\zeta(z,a). \tag{2.3}$$

The interest of such formulas has been manifest since some years ago, and actually a couple of attempts had been made by some authors to solve the problem, but they did not turn out to be completely successful. It is rather clear from the very beginning that, for the general case (2.3), it is not possible to obtain an expression so simple as the one derived in Ref. [95] for (2.1) (see (2.8) below).

We will not repeat here the detailed derivation of the asymptotic expansion corresponding to (2.1), given in Ref. [95]. Starting from Hermite's integral representation of the Hurwitz zeta function $\zeta(z,a)$ (in the future we will omit the subindex H, as is normal practice)

$$\zeta(z,a) = \frac{a^{-z}}{2} + \frac{a^{1-z}}{z-1} + 2\int_0^{\infty}(t^2+a^2)^{-z/2}\sin(z\tan^{-1}(t/a))\frac{dt}{e^{2\pi t}-1}, \tag{2.4}$$

one easily gets

$$\begin{aligned}
\zeta'(z,a) = -\frac{a^{-z}}{2}\ln a - \frac{a^{1-z}}{z-1}\ln a - \frac{a^{1-z}}{(z-1)^2} \\
+ 2\int_0^{\infty}(t^2+a^2)^{-z/2}\cos(z\tan^{-1}(t/a))\tan^{-1}(t/a)\frac{dt}{e^{2\pi t}-1} \\
- \int_0^{\infty}(t^2+a^2)^{-z/2}\sin(z\tan^{-1}(t/a))\ln(t^2+a^2)\frac{dt}{e^{2\pi t}-1}.
\end{aligned} \tag{2.5}$$

We invite the reader to read Ref. [95] for more details of the mathematical procedure employed, which is similar to the ordinary one derived from Watson's lemma and Laplace's method [84, 99] and is therefore quite a conventional one for the obtaintion of asymptotic expansions. In particular, the functions appearing in the integrands are replaced by their power series expansions near $t = 0$, e.g.

$$\tan^{-1}\left(\frac{t}{a}\right) = \sum_{j=0}^{\infty} \frac{(-1)^j}{2j+1}\left(\frac{t}{a}\right)^{2j+1},$$

$$\ln(t^2 + a^2) = 2\ln a + \sum_{j=0}^{\infty} \frac{(-1)^{j-1}}{j}\left(\frac{t}{a}\right)^{2j},$$

(2.6)

and one then checks with the residuum terms that the series that are formed by integrating term by term verify the condition of asymptoticity. Alternatively, another procedure can be employed that leads to the same result, namely repeated integration by parts.

In any way, the final result turns out to be the same [97] that one would obtain by naive derivation term by term of the asymptotic series corresponding to the Hurwitz zeta function (2.2)

$$\zeta(z+1, a) = \frac{1}{z}a^{-z} + \frac{1}{2}a^{-z-1} + \frac{1}{z}\Sigma_0(z, a),$$

$$\Sigma_0(z, a) \equiv \sum_{k=2}^{\infty} \frac{B_k}{k!}(z)_k\, a^{-z-k},$$

(2.7)

$$(z)_k \equiv z(z+1)\cdots(z+k-1) = \frac{\Gamma(z+k)}{\Gamma(z)}.$$

Here $(z)_k$ is Pochhammer's symbol (the rising factorial function) and the B_k are Bernoulli's numbers. The asymptotic series corresponding to $\zeta'(z+1, a)$ can be expressed as

$$\zeta'(z+1, a) = -\left(\frac{1}{z} + \ln a\right)\zeta(z+1, a) + \frac{1}{2z}a^{-z-1} + \frac{1}{z}\Sigma_1(z, a),$$

$$\Sigma_1(z, a) \equiv \sum_{k=2}^{\infty} B_k \sum_{j=0}^{k-1} \frac{(z)_j}{j!(k-j)}a^{-z-k}.$$

(2.8)

Notice that this is not a trivial result since—as is well known—term by term derivation of an asymptotic series is controlled by a Tauberian theorem (and *not* by an abelian one). These expansions are valid for large $|a|$ and $|\operatorname{Arg} a| < \pi$. In particular, when $z = -n$, $n \in \mathbb{N}$, the above expressions reduce to [100]

$$\zeta(1-n, a) = -\frac{1}{n}B_n(a),$$

(2.9)

where $B_n(a)$ is the Bernoulli polynomial of degree n, and to

$$\zeta'(1-n,a) = \frac{1}{n}\left(\ln a - \frac{1}{n}\right)B_n(a) - \frac{1}{2n}a^{n-1}$$

$$-\frac{1}{n}\sum_{k=2}^{n}B_k\sum_{j=0}^{k-1}\binom{n}{j}\frac{(-1)^j}{k-j}a^{n-k}$$

$$+(-1)^{n-1}(n-1)!\sum_{k=n+1}^{\infty}\frac{B_k}{k(k-1)\cdots(k-n)}a^{n-k}, \quad (2.10)$$

respectively.

Now, starting again from Hermite's integral representation (2.4) and repeating, for ζ'', the same procedure used for ζ', in particular, the replacements (2.6) or integration by parts (quite involved), we arrive to the following expression for the second derivative

$$\zeta''(z+1,a) = -2\left(\frac{1}{z}+\ln a\right)\zeta'(z+1,a)$$

$$-\left(2\frac{\ln a}{z}+\ln^2 a\right)\zeta(z+1,a) + \frac{1}{z}\Sigma_2(z,a), \quad (2.11)$$

$$\Sigma_2(z,a) \equiv \sum_{k=2}^{\infty}B_k\sum_{j=0}^{k-1}\frac{1}{k-j}\sum_{h=0}^{j-1}\frac{(z)_h}{h!(j-h)}a^{-z-k}.$$

This is, term by term, the same result that one would have obtained by naive derivation of the preceding asymptotic series. Actually, the alternative procedure (namely that of partial integration)—which was already discussed in [95]—proves to be here the most convenient one in order to exhibit the asymptotic character of the series (2.11).

With some additional effort, the following operational recurrence can be found, in general [98]

$$\frac{\partial^m}{\partial z^m}\zeta(z+1,a)$$

$$= -\left[\left(\frac{\partial}{\partial z}+\ln a\right)^m - \left(\frac{\partial}{\partial z}\right)^m + \frac{m}{z}\left(\frac{\partial}{\partial z}+\ln a\right)^{m-1}\right]\zeta(z+1,a)$$

$$+\frac{1}{z}\Sigma_m(z,a), \quad (2.12)$$

being

$$\Sigma_m(z,a) \equiv \sum_{k=2}^{\infty}B_k\sum_{j_1=0}^{k-1}\frac{1}{k-j_1}\sum_{j_2=0}^{j_1-1}\frac{1}{j_1-j_2}\cdots\sum_{j_m=0}^{j_{m-1}-1}\frac{(z)_{j_m}}{j_m!(j_{m-1}-j_m)}a^{-z-k},$$

$$(2.13)$$

for large $|a|$ and $|\operatorname{Arg} a| < \pi$. Using the operational iteration (2.12), the following general recurrence is obtained, which yields the asymptotic expansion for any derivative of the Hurwitz zeta function in terms of the asymptotic expansion corresponding to the derivatives of lower order:

$$\zeta^{(m)}(z+1,a) = -\sum_{j=1}^{m} \binom{m}{j}\left(\frac{j}{z} + \ln a\right)\ln^{j-1} a \zeta^{(m-j)}(z+1,a) + \frac{1}{z}\Sigma_m(z,a),$$

(2.14)

$\Sigma_m(z,a)$ being given by (2.13).

If we restrict ourselves to the particular values $z = -n$, $n \in \mathbb{N}$, we obtain the more simple expression

$$\zeta^{(m)}(1-n,a) = \sum_{j=1}^{m} \binom{m}{j}\left(\frac{j}{n} - \ln a\right)\ln^{j-1} a \zeta^{(m-j)}(1-n,a) - \frac{1}{n}\Sigma_m(-n,a),$$

(2.15)

where now

$$\Sigma_m(-n,a) = \sum_{k=2}^{\infty} B_k \sum_{j_1=0}^{k-1}\frac{1}{k-j_1}\sum_{j_2=0}^{j_1-1}\frac{1}{j_1-j_2}\cdots\sum_{j_m=0}^{\mu_m}\binom{n}{j_m}\frac{(-1)^j}{j_{m-1}-j_m}a^{n-k}, \quad (2.16)$$

being $\mu_m = \min(n, j_{m-1} - 1)$.

(2.12) to (2.16) constitute the main results of this section. Even if they do not provide a general explicit asymptotic expression for any derivative of the Hurwtiz zeta function but any of such asymptotic series can immediately be found by solving the very simple recurrences (2.14) or (2.15), starting from (2.7) and (2.9), and (2.8) and (2.10), respectively. These expressions are very appropriate for numerical and analytical explicit calculations, in connection with the computational software packets commonly available.

To prove this statement, let us obtain the asymptotic series for the second derivative, at non-positive integer values of z. It is given by

$$\zeta''(1-n,a) = \left(-\frac{2}{n^2} + \frac{2\ln a}{n} - \ln^2 a\right)\frac{B_n(a)}{n} - \left(\frac{1}{n} - \ln a\right)\frac{a^{n-1}}{n}$$

$$- \frac{1}{n}\Sigma_2(-n,a) - \frac{2}{n}\left(\frac{1}{n} - \ln a\right)\Sigma_1(-n,a), \quad (2.17)$$

where $B_n(a)$ is the Bernoulli polynomial of degree n, and

$$\Sigma_2(-n,a) = \sum_{k=2}^{n} B_k \sum_{j=0}^{k-1}\frac{1}{k-j}\sum_{h=0}^{j-1}\binom{n}{h}\frac{(-1)^h}{j-h}a^{n-k}$$

$$+ \sum_{k=n+1}^{\infty} B_k \sum_{j=0}^{n}\frac{1}{k-j}\sum_{h=0}^{j-1}\binom{n}{h}\frac{(-1)^h}{j-h}a^{n-k}$$

$$+ (-1)^n n! \sum_{k=n+1}^{\infty} B_k \sum_{j=n+1}^{k-1} \frac{1}{(k-j)j(j-1)\cdots(j-n)} a^{n-k} \quad (2.18)$$

and

$$\Sigma_1(-n, a) = \sum_{k=2}^{n} B_k \sum_{j=0}^{k-1} \binom{n}{j} \frac{(-1)^j}{k-j} a^{n-k}$$

$$+ (-1)^n n! \sum_{k=n+1}^{\infty} \frac{B_k}{k(k-1)\cdots(k-n)} a^{n-k}. \quad (2.19)$$

It is also clear enough that these expressions for the asymptotic series are well suited for practical purposes. We have used *Mathematica* in a conventional workstation in order to obtain a number of leading terms of the above series, for different values of n. This can be done in less than a minute. Below there is a list of the first few results obtained (we will not bother the reader with the full sample):

$$\zeta''(0, a) = -a\left(\ln^2 a - 2\ln a + 1\right) + \frac{1}{2}\ln^2 a - \frac{a^{-1}}{6}\ln a$$

$$+ \frac{a^{-3}}{60}\left(\frac{\ln a}{3} - \frac{1}{2}\right) - a^{-5}\left(\frac{\ln a}{630} - \frac{4}{1209}\right) + \cdots, \quad (2.20)$$

$$\zeta''(-1, a) = -\frac{a^2}{4}\left(2\ln^2 a - 2\ln a + 1\right) + \frac{a}{2}\ln^2 a + \frac{1}{12}\left(\ln^2 a - 2\ln a\right)$$

$$- \frac{a^{-2}}{360}\ln a + \frac{a^{-4}}{15120}(6\ln a - 5) + \cdots, \quad (2.21)$$

$$\zeta''(-2, a) = -\frac{a^3}{27}\left(9\ln^2 a - 6\ln a + 2\right) + \frac{a^2}{2}\ln^2 a - \frac{a}{6}\left(\ln^2 a - \ln a\right)$$

$$+ \frac{a^{-1}}{60}\left(\frac{\ln a}{3} + \frac{1}{2}\right) - \frac{a^{-3}}{3780}\ln a$$

$$+ \frac{a^{-5}}{1800}\left(\frac{\ln a}{7} - \frac{1}{12}\right) + \cdots, \quad (2.22)$$

and

$$\zeta''(-3, a) = -\frac{a^4}{32}\left(8\ln^2 a - 4\ln a + 1\right) + \frac{a^3}{2}\ln^2 a + \frac{a^2}{12}\left(3\ln^2 a + 2\ln a\right)$$

$$+ \frac{1}{60}\left(\frac{\ln^2 a}{2} + \frac{11\ln a}{6} + 1\right)$$

$$+ \frac{a^{-2}}{504}\left(\frac{\ln a}{5} + \frac{1}{6}\right) - \frac{a^{-4}\ln a}{16800} + \cdots. \quad (2.23)$$

Actually, one reaches values of n as high as $n = 20$ very quickly, what proves that the above recurrent expressions, (2.13) to (2.16), are in fact very efficient for practical applications.

2.2 The Zeta-Function Regularization Theorem

As advanced before, the zeta-function regularization procedure is a quite useful regularization tool in quantum field theory. A keystone of the method is the zeta function regularization theorem. In this section, the theorem will be illustrated while addressing the practical question of the regularization of multi-series of the general type

$$\sum_{n_1,\ldots,n_N} \left[a_1(n_1 + c_1)^{\alpha_1} + \cdots + a_N(n_N + c_N)^{\alpha_N} + c \right]^{-s}, \qquad (2.24)$$

with $a_1, \ldots, a_N, \alpha_1, \ldots, \alpha_N > 0$, $c_1, \ldots, c_N$ arbitrary reals and $c \geq 0$. When $c_1 = \cdots = c_N = 0$ the term with $n_1 = \cdots = n_N = 0$ must be suppressed from the sum (which is then usually denoted by Σ'). Only the most simple cases have been properly studied in the literature (e.g., $a_1 = \cdots = a_N$, $c_1 = \cdots = c_N = 0$ or $\pm 1/2$, $\alpha_1 = \cdots = \alpha_N = 1, 2$, $c = 0$, etc.). The zeta function regularization theorem in its most general form leads to an asymptotic expansion valid for arbitrary a's and α's, which is very convenient for numerical computations. In particular, useful expressions can be derived from it for the analytical continuation of Riemann, Hurwitz and Epstein zeta functions and their generalizations (see Chap. 4), and for their asymptotic expansions—including those of derivatives and integrals. Physical applications of the zeta-regularization procedure include the proper definition of the vacuum energy, the Casimir effect, spontaneous compactification in quantum gravity, stability analysis of strings and membranes, etc., and embrace also very recent experiments of solid state and condensed matter physics employing liquid helium (those will be described in the following chapters).

The method of zeta-function regularization has a rather long history. There are precedents in the use of Riemann and Epstein zeta functions as summation (i.e., regularization) procedures in the late sixties [44, 45, 47]. However, the zeta-function regularization method as such was introduced in the middle seventies [23, 24, 39–41]. The paper by Hawking [24] (of 1977) is generally considered as the first systematic description of the zeta function procedure as a useful technique in physics for providing the finite values corresponding to path integrals over fields in curved backgrounds and for the evaluation of determinants of quadratic differential operators (see, however, the other references mentioned, in which the method had already been applied before). The calculation of determinants of differential operators is a basic, multipurpose need in theoretical physics and in several branches of mathematics (such as analysis and number theory).

In the last 15 years the zeta-regularization procedure has been used more and more by the leading physicists and mathematicians and we can nowadays say that it is a basic procedure of quantum field theory. At the beginning the method was rather simple minded, but nowadays it comprises a whole set of different techniques, of increasing difficulty, to treat the several degrees of complexity of the physical (and corresponding mathematical) problems to be solved.

The list of people who have been dealing with zeta functions at one instance or other would be just non-ending. Maybe Al Actor is one of the persons that have devoted more years to this subject (at least among those of the mathematical-physicists squad). According to Actor himself [53], a milestone in the field of regularization of discrete sums of the general form (2.24) has been the proof of the so-called *zeta-function regularization theorem*. In its final formulation, it is the result of hard work of A.A. Actor, H.A. Weldon, A. Romeo, and the author [48, 101, 102]. The uses and applications of the theorem in its most general form [48]—for discrete series of the type (2.24)—are very far reaching. In particular it leads to asymptotic expansions, valid for arbitrary a's and α's, of the multi-series of this general kind, which are well suited for numerical computations. These expansions are unchallenged in its usefulness for such purposes. They will be presented later in this chapter.

The zeta function regularization theorem provides a method for the computation of expressions like (2.24)—and even more involved ones—valid for Re(s) big enough, in terms of their analytic (usually meromorphic) continuation to other values of s. In the zeta-function procedure they are given as combinations of the ordinary Riemann and Hurwitz zeta-functions.

A very simple case corresponds to the Hamiltonian zeta-function $\zeta(s) \equiv \sum_i E_i^{-s}$, with E_i eigenvalues of H [103, 104]. For a system of N non-interacting harmonic oscillators, one has $\alpha_j = 1$, $j = 1, 2, \ldots, N$, and the a_j are the corresponding eigenfrequencies ω_j. Another interesting case is partial toroidal compactification (spacetime $\mathbb{T}^p \times \mathbb{R}^{q+1}$). Then $\alpha_j = 2$ and, usually, $c_j = 0, \pm 1/2$. One is thus led to the Epstein zeta-functions [105–107]

$$
Z_N(s) = \sum_{n_1,\ldots,n_N=-\infty}^{\infty}{}' \left(n_1^2 + \cdots + n_N^2\right)^{-s},
$$

$$
Y_N(s) = \sum_{n_1,\ldots,n_N=-\infty}^{\infty}{}' \left[\left(n_1 + \frac{1}{2}\right)^2 + \cdots + \left(n_N + \frac{1}{2}\right)^2\right]^{-s}
$$

(2.25)

(remember that the prime means omission of the term $n_1 = \cdots = n_N = 0$). Other powers α_j appear when one deals with the spherical compactification (spacetime $\mathbb{S}^p \times \mathbb{R}^{q+1}$) and with more involved ones arising, e.g., in superstring theory and their membrane and p-brane generalizations [10, 11, 13, 14]. Hence, the general expression (2.24). The only precedents in the literature (to our knowledge) of this kind of evaluations have been restricted to few special cases other than $a_1 = \cdots = a_N$ and $c_1 = \cdots = c_N = 0$. Very famous is the expression due to Hardy [17], a particular case of our final formula.

2.2.1 The Theorem (Special Form)

An interesting result concerning the interchange of the order of summation of the infinite series appearing in zeta-function regularization is due to Weldon [102]. His investigation originated in some difficulties which appeared in a paper by Actor [101] when he tried to obtain the value of the thermodynamical potential corresponding to a relativistic Bose gas by using the zeta-function regularization procedure. Unfortunately, Weldon's proof had its own limitations, and the statements in [102] concerning the extent of its validity were actually not right. This is quite easy to check in some particular cases, and was stressed in [108].

Let us briefly summarize the nice proof due to Weldon of the validity of the zeta-function regularization procedure [102] and point out its shortcomings. Using the same notation as in [102], let us consider the four series

$$S_F = \sum_{m=1}^{\infty} \frac{(-1)^{m+1}}{m^{s+1}} \sum_{a=0}^{\infty} m^a f(a), \tag{2.26}$$

$$S_B = \sum_{m=1}^{\infty} \frac{1}{m^{s+1}} \sum_{a=0}^{\infty} m^a f(a), \tag{2.27}$$

$$S_{AF} = \sum_{m=1}^{\infty} \frac{(-1)^{m+1}}{m^{s+1}} \sum_{a=0}^{\infty} (-1)^a m^a f(a), \tag{2.28}$$

$$S_{AB} = \sum_{m=1}^{\infty} \frac{1}{m^{s+1}} \sum_{a=0}^{\infty} (-1)^a m^a f(a), \tag{2.29}$$

where $f(a) \geq 0$ for positive integer a. They are assumed to be convergent, as they stand. The idea of the zeta-function regularization procedure begins with the interchange of the order of the summation of the two infinite series involved in each case.

Theorem 1 Let $f(a)$ be defined in the complex a-plane, satisfying:

1. The function $f(a)$ is regular for $\mathrm{Re}\, a \geq 0$.
2. Either
   (a) in the case of (2.26) and (2.27), $am^a f(a) \to 0$, as $|a| \to \infty$, for $\mathrm{Re}\, a \geq 0$ and fixed m;
   (b) in the case of (2.28) and (2.29), $am^a f(a)e^{-\pi|\mathrm{Im}\, a|} \to 0$, as $|a| \to \infty$, for $\mathrm{Re}\, a \geq 0$ and fixed m.

Then it turns out that, in the fermionic cases, (2.26) and (2.28), one can naively interchange the order of the summations, to get

$$S_F = \sum_{a=0}^{\infty} \eta(s+1-a)f(a), \qquad S_{AF} = \sum_{a=0}^{\infty} (-1)^a \eta(s+1-a)f(a), \tag{2.30}$$

while in the bosonic cases, (2.27) and (2.29), one obtains the additional contributions

$$S_B = \sum_{a=0}^{\infty} \zeta(s+1-a)f(a) - \pi \,\mathrm{ctg}(\pi s)f(s), \quad s \notin \mathbb{N},$$

$$S_B = \sum_{\substack{a=0 \\ a \neq s}}^{\infty} \zeta(s+1-a)f(a) + \gamma f(s) - f'(s), \quad s \in \mathbb{N},$$

(2.31)

and

$$S_{AB} = \sum_{a=0}^{\infty} (-1)^a \zeta(s+1-a)f(a) - \pi \csc(\pi s)f(s), \quad s \notin \mathbb{N},$$

$$S_{AB} = \sum_{\substack{a=0 \\ a \neq s}}^{\infty} (-1)^a \zeta(s+1-a)f(a) + (-1)^s [\gamma f(s) - f'(s)], \quad s \in \mathbb{N},$$

(2.32)

respectively. Here $\zeta(s)$ and $\eta(s)$ are the Riemann ordinary and alternating zeta functions:

$$\zeta(s) = \sum_{m=1}^{\infty} m^{-s}, \quad \mathrm{Re}\,s > 1,$$

$$\eta(s) = \sum_{m=1}^{\infty} (-1)^{m+1} m^{-s}, \quad \mathrm{Re}\,s > 0,$$

(2.33)

$$\eta(s) = (1 - 2^{1-s})\zeta(s),$$

γ is Euler–Mascheroni's constant, and $f'(s)$ means derivative of f with respect to s.

The proof of the preceding theorem proceeds by integration in the complex a-plane. One writes (2.26) to (2.29) under the form of contour integrals

$$S_F = \sum_{m=1}^{\infty} \frac{(-1)^{m+1}}{m^{s+1}} \oint_C \frac{da}{2i} m^a f(a) \cot(\pi a),$$

(2.34)

$$S_B = \sum_{m=1}^{\infty} \frac{1}{m^{s+1}} \oint_C \frac{da}{2i} m^a f(a) \cot(\pi a),$$

(2.35)

$$S_{AF} = \sum_{m=1}^{\infty} \frac{(-1)^{m+1}}{m^{s+1}} \oint_C \frac{da}{2i} m^a f(a) \csc(\pi a),$$

(2.36)

$$S_{AB} = \sum_{m=1}^{\infty} \frac{1}{m^{s+1}} \oint_C \frac{da}{2i} m^a f(a) \csc(\pi a),$$

(2.37)

Fig. 1 The *closed contour C* (that one should always follow counterclockwise, i.e., with the positive orientation) consists of the *straight line* $\operatorname{Re} a = a_0$, $0 < a_0 < 1$, and of the *semicircumference* 'at infinity' on its right, K

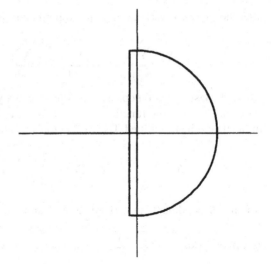

where C is the closed contour defined by the straight line $\operatorname{Re} a = -a_0$—for fixed a_0 such that $0 < a_0 < 1$—and by the semicircumference at infinity on the right (see Fig. 1). The contribution from the semicircumference is zero in every case, due to the asymptotic behavior of $f(a)$ and, as long as $\operatorname{Re} s > -1$, integration extended to the line $\operatorname{Re} a = -a_0$ can be interchanged with the remaining summation over m. The final step is to close the contour C again with the semicircumference at infinity. In the cases (2.35) and (2.37) there appears an additional contribution from the pole of the zeta function $\zeta(s + 1 - a)$ at $a = s$. On the contrary, in the cases (2.34) and (2.36) the alternating zeta function $\eta(s + 1 - a)$ has no pole in the region enclosed by C. All the steps in this procedure are quite simple and one obtains (2.30) to (2.32).

However it was further explicitly stated by Weldon in [102] that the results for the alternating fermionic and for the alternating bosonic cases, S_{AF} and S_{AB}, respectively, could be naively extended to the following types of series

$$S_{AF}^{(N)} = \sum_{m=1}^{\infty} \frac{(-1)^{m+1}}{m^{s+1}} \sum_{a=0}^{\infty} (-1)^a m^{Na} f(a),$$

$$S_{AB}^{(N)} = \sum_{m=1}^{\infty} \frac{1}{m^{s+1}} \sum_{a=0}^{\infty} (-1)^a m^{Na} f(a), \tag{2.38}$$

with N any positive integer. By going through the same proof once more, he just obtained a trivial modification of the above results. That this generalization of (2.30) and (2.32) and for any positive integer N is in error is easy to check. In particular, it was noticed by Actor in [108]. As a clear example, let us study the simplest case after the (only correct) one $N = 1$ (explicitly considered in [102]), i.e. $N = 2$. Let

$$S \equiv \sum_{m=1}^{\infty} e^{-m^2} = \sum_{m=1}^{\infty} \frac{1}{m^{s+1}} \sum_{a=0}^{\infty} (-1)^a \frac{m^{2a}}{a!} \bigg|_{s=-1}, \qquad (2.39)$$

where the last operation consists in doing the analytic continuation of the resulting series to $s = -1$. The function $f(a)$ is here $f(a) = \frac{1}{\Gamma(a+1)}$ and all the hypotheses of the theorem are fulfilled. Use of Weldon's formula gives

$$S = \sum_{a=0}^{\infty} \frac{(-1)^a}{a!} \zeta(-2a) - \frac{\frac{\pi}{2} \csc(-\frac{\pi}{2})}{\Gamma(1-\frac{1}{2})} = -\frac{1}{2} + \frac{\sqrt{\pi}}{2}, \qquad (2.40)$$

which is false, though numerically almost undetectable, because

$$S = 0.3863186, \quad \frac{\sqrt{\pi}-1}{2} = 0.3862269, \quad \Delta \equiv \frac{\sqrt{\pi}-1}{2} - S = -9.17 \times 10^{-5}. \qquad (2.41)$$

Going on to $N = 2, 3, 4, \ldots$, it is not difficult to see that, if N is constrained to be a positive integer, Weldon's formula is true only for $N = 1$ ((2.30) and (2.32)).

As the author managed to demonstrate in [48], the step which fails to be correct in Weldon's proof for general N is the last one, namely, even if the asymptotic behavior (2b) of the function $f(a)$ allows us to suppress the contribution from the curved contour in the second step, this will be no longer true when we try to close again the circuit C in the last step. *There is in fact a contribution coming from the integral of $\zeta(s+1-Na)f(a)$ over the semicircumference at infinity* (due to the asymptotic behavior of the zeta-function). And this is so whatever it be the value we choose for s. The study of the asymptotic behavior of $\zeta(s+1-Na)$ immediately distinguishes the case $N \leq 1$ from $N > 1$. It is, however, misleading in some sense, because the fact that the zeta-function diverges for $N > 1$ does *not* necessarily mean that the contour actually provides a non-zero contribution invalidating Weldon's proof (that had been conjectured by Actor, at a first instance). Things must be done with great care due to the presence of highly oscillating factors.

Let us restrict the argument to the case $f(a) = \frac{1}{\Gamma(a+1)}$. This is enough for many applications and the generalization to other situations proceeds by analogy. In this case, the fact that the poles of Γ are the non-positive integers and a suitable application of the zeta function reflection formula allow us to write the additional contribution as a contour integral over a curved path in the complex left half-plane. Besides, by using the relation

$$\Gamma\left(\frac{z}{2}\right) \zeta(z) = \int_0^{\infty} dt \, t^{z/2-1} S_2(t), \quad \mathrm{Re}\, z > 0, \qquad (2.42)$$

where

$$S_\alpha(t) \equiv \sum_{m=1}^{\infty} e^{-m^\alpha t}, \qquad (2.43)$$

and owing to the behavior of the complex function $\Gamma(z)$ which has simple poles at $z = -n$ for $n = 0, 1, 2, \ldots$, with residues

$$\text{Res}_{z=-n} \Gamma(z) = \frac{(-1)^n}{n!}, \tag{2.44}$$

and with the aid of

$$\Gamma(z)\Gamma(1 - z) = \pi \csc(\pi z), \tag{2.45}$$

we can write

$$S_{AB}^{(\alpha)} \equiv \sum_{m=1}^{\infty} \frac{1}{m^{s+1}} \sum_{a=0}^{\infty} (-1)^a \frac{m^{\alpha a}}{\Gamma(a + 1)}, \quad \alpha \in \mathbb{R}, \tag{2.46}$$

as

$$S_{AB}^{(\alpha)} \equiv \sum_{m=1}^{\infty} \frac{1}{m^{s+1}} \oint_{\bar{C}} \frac{da}{2\pi i} m^{-\alpha a} \Gamma(a), \tag{2.47}$$

where now the contour $\bar{C}$ consists of the line $\text{Re}\, a = a_0$, with a_0 fixed, $0 < a_0 < 1$, and of the semicircumference at infinity on the left. For $s = -1$,

$$S_{AB}^{(\alpha)}(s = -1) = \sum_{m=1}^{\infty} \sum_{a=0}^{\infty} (-1)^a \frac{m^{\alpha a}}{a!} = \sum_{m=1}^{\infty} e^{-m^\alpha} = S_\alpha(1). \tag{2.48}$$

Finally, after correctly making the last step in the above proof, we end up with

$$S_{AB}^{(\alpha)} = \sum_{a=0}^{\infty} \frac{(-1)^a}{a!} \zeta(s + 1 - \alpha a) + \frac{1}{\alpha} \Gamma\left(-\frac{s}{\alpha}\right) - \Delta_{AB}^{(\alpha)}, \quad \frac{s}{\alpha} \notin \mathbb{N}, \tag{2.49}$$

$$S_{AB}^{(\alpha)} = \sum_{\substack{a=0 \\ a \neq s/\alpha}}^{\infty} \frac{(-1)^a}{a!} \zeta(s + 1 - \alpha a) + (-1)^{\frac{s}{\alpha}} \left[\frac{\gamma}{\Gamma(\frac{s}{\alpha} + 1)} + \frac{\Gamma'(\frac{s}{\alpha} + 1)}{\alpha \Gamma^2(\frac{s}{\alpha} + 1)} \right]$$

$$- \Delta_{AB}^{(\alpha)}, \quad \frac{s}{\alpha} \in \mathbb{N}, \tag{2.50}$$

where $\Delta_{AB}^{(\alpha)}$ is the contribution of the curved part K of the contour $\bar{C}$—which consists now of the line $\text{Re}\, a = a_0$, for fixed a_0 such that $0 < a_0 < 1$ and by the semicircumference at infinity on the left (see Fig. 2)

$$\Delta_{AB}^{(\alpha)} \equiv \int_K \frac{da}{2\pi i} \zeta(s + 1 + \alpha a) \Gamma(a). \tag{2.51}$$

This contribution is non-zero for any value of s. We can check that it actually provides the term missing from (2.40). Before proceeding with the actual calculation of (2.51), one can, as an illustrating exercise, close back the contour on the right instead of on the left, and see that the same series is obtained (a desired identity!).

Fig. 2 The *closed contour* $\overline{C}$ (also counterclockwise) consists of the *straight line* $\mathrm{Re}\, a = a_0$, $0 < a_0 < 1$, and of the *semicircumference* 'at infinity' on its left

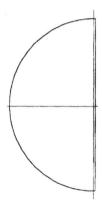

Coming back to (2.51) and doing the same for $s = -1$ and $\alpha = 2$, we must use first the reflection formula for the zeta function

$$\Gamma\left(\frac{z}{2}\right)\zeta(z) = \pi^{z-1/2}\Gamma\left(\frac{1-z}{2}\right)\zeta(1-z), \tag{2.52}$$

what yields

$$\Delta_{AB}^{(2)}(s = -1) = \int_K \frac{da}{2i\sqrt{\pi}} \int_0^\infty dt\, t^{-a-1/2} S_2(\pi^2 t) = -\sqrt{\pi}\, S_2(\pi^2), \tag{2.53}$$

that is

$$S_2(1) = -\frac{1}{2} + \frac{\sqrt{\pi}}{2} + \sqrt{\pi}\, S_2(\pi^2). \tag{2.54}$$

This result happens to be just a particular case of Jacobi's theta function identity

$$\theta_3(z, \tau) = \tau^{-1/2} e^{\pi z^2/\tau} \theta_3\left(\frac{z}{i\tau}, \frac{1}{\tau}\right), \tag{2.55}$$

θ_3 being the elliptic function

$$\theta_3(z, \tau) = \sum_{n=-\infty}^{\infty} e^{-\pi n^2 \tau + 2\pi nz}, \quad z \in \mathbb{C}, \ \tau \in \mathbb{R}^+. \tag{2.56}$$

Notice that $S_2(\pi t) = \frac{1}{2}[\theta(0, t) - 1]$. (2.53) is an exact expression. Once more, we observe that the contribution of the contour provides, in fact, the missing term.

Let us now again consider (2.44) for general α and, $s = -1$. (2.42) and (2.43) read, in this case,

$$\Gamma(z)\zeta(\alpha z) = \int_0^\infty dt\, t^{z-1} S_\alpha(t), \tag{2.57}$$

$S_\alpha(t)$ being the function given in (2.48). No simple reflection formula like (2.52) exists for $\alpha \neq 2$. We have, instead,

$$\zeta(\alpha z) = \frac{2\Gamma(1 - \alpha z)}{(2\pi)^{1-\alpha z}} \sin\left(\frac{\pi \alpha z}{2}\right) \zeta(1 - \alpha z), \tag{2.58}$$

and we get

$$S_\alpha \equiv S_\alpha(1) = \sum_{m=1}^{\infty} e^{-m^\alpha} = \sum_{a=0}^{\infty} \frac{(-1)^a}{a!} \zeta(-\alpha a) + \frac{1}{\alpha}\Gamma\left(\frac{1}{\alpha}\right) - \Delta_\alpha, \tag{2.59}$$

being the contribution of the contour

$$\Delta_\alpha = \int_K \frac{da}{2\pi i} \zeta(\alpha a)\Gamma(a). \tag{2.60}$$

Putting everything together, we have proven the following

Theorem 2 (Zeta function regularization theorem, particular case) *Under the hypothesis* (1), (2a) *and* (2b) *above, we have that*:

1. *For* $-\infty < \alpha < 2$, *the contribution of the semicircumference at infinity is zero, i.e.*

$$\Delta_\alpha = 0, \quad \alpha < 2. \tag{2.61}$$

2. *For* $\alpha = 2$, *the contribution of the semicircumference at infinity is given by*

$$\Delta_2 = -\sqrt{\pi}\, S_2(\pi^2). \tag{2.62}$$

The result for $\alpha \leq 1$ was known already and constitutes Weldon's proof of zeta-function regularization. The result for $\alpha = 2$ is due to the author. It shows very clearly that, on the contrary, the statements in [102] about the validity of the proof for any positive integer α were false, the reason being that the semicircumference at infinity does *not* yield a zero contribution. It was precisely the last step of the proof in [102] that was wrong. The fact that the numerical value of Δ_α is so small (it can be thought of as an infinitesimal correction, see (2.62)) as compared with the rest of the terms in (2.49) and (2.50) gives sense to the whole procedure of zeta-function regularization.

However, this is strictly true only for small α. For large α, Δ_α ceases to be an infinitesimal contribution. Actually, in the case considered,

$$\Delta_\alpha = 0, \quad \alpha < 2;$$
$$\Delta_2 = 9.17 \times 10^{-5}; \quad \Delta_4 = 0.04; \quad \Delta_6 = 0.07; \tag{2.63}$$
$$\Delta_\alpha \to 0.13, \quad \alpha \to \infty,$$

which represent, respectively, contributions of the 0%, 0.02%, 11%, 19%, and 36% to the final value of $S_\alpha(1)$. A more precise statement on this point—together with a substantial extension of the theorem to general situations—will be given in the next chapter.

2.3 Immediate Application of the Theorem

The most interesting and simple of the new cases is when $0 < c < 1$, e.g.

$$S_c \equiv S_c^{(2)}(-1) = \sum_{m=0}^{\infty} e^{-(m+c)^2}. \tag{2.64}$$

It can be expressed in terms of Hurwitz zeta-functions, as

$$S_c = \sum_{m=0}^{\infty} \frac{(-1)^m}{m!} \zeta(-2m, c) + \frac{\sqrt{\pi}}{2} + \sqrt{\pi} \cos(2\pi c) S(\pi^2), \tag{2.65}$$

where $S(t) \equiv \sum_{m=0}^{\infty} e^{-tm^2}$. For $c \neq 0, 1/2$, this series is asymptotic.

1. Particular case $c = 1$: we recover the known equality (a special case of Jacobi's one)

$$S(1) = \frac{\sqrt{\pi} - 1}{2} + \sqrt{\pi} S(\pi^2). \tag{2.66}$$

2. For $c = 1/2$ we have $\zeta(-2m, 1/2) = 0$, $m = 0, 1, 2, \ldots$, and the result

$$\sum_{m=0}^{\infty} \exp\left[-\left(m + \frac{1}{2}\right)^2\right] = \frac{\sqrt{\pi}}{2} - \sqrt{\pi} \sum_{m=1}^{\infty} \exp(-m^2 \pi^2) \tag{2.67}$$

permits us to obtain the value of the series with 10^{-10} accuracy, with just two terms

$$\sum_{m=0}^{\infty} \exp\left[-\left(m + \frac{1}{2}\right)^2\right] = \frac{\sqrt{\pi}}{2} - \sqrt{\pi} e^{-\pi^2} + \mathcal{O}(10^{-10}). \tag{2.68}$$

3. For $c = 0$ we obtain the previous result (2.54)

$$\sum_{m=0}^{\infty} e^{-m^2} = \frac{1}{2} + \frac{\sqrt{\pi}}{2} + \sqrt{\pi} S(\pi^2). \tag{2.69}$$

4. For $c = 1/4$, we have

$$\sum_{m=0}^{\infty} \exp\left[-\left(m + \frac{1}{4}\right)^2\right] \sim \frac{\sqrt{\pi}}{2} + \sum_{m=0}^{\infty} \frac{(-1)^m}{m!} \zeta\left(-2m, \frac{1}{4}\right). \tag{2.70}$$

The series is now asymptotic. It stabilizes between the 8th and the 12th summand and provides its best value (with $\simeq 10^{-7}$ accuracy) when we add the 10 first terms (optimal truncation of the asymptotic series).

5. For $c = 1/3$ and $c = 1/6$,

$$\sum_{m=0}^{\infty} \exp\left[-\left(m + \frac{1}{3j}\right)^2\right] \sim \frac{\sqrt{\pi}}{2} + \sum_{m=0}^{\infty} \frac{(-1)^m}{m!} \zeta\left(-2m, \frac{1}{3j}\right)$$

$$+ (-1)^j \frac{\sqrt{\pi}}{2} \sum_{m=1}^{\infty} \exp(-m^2\pi^2), \quad j = 1, 2. \quad (2.71)$$

6. Some more relations are

$$\sum_{m=-\infty}^{\infty} \exp\left[-(m+c)^2\right] \sim \sqrt{\pi} + 2\sqrt{\pi} \cos(2\pi c) S(\pi^2),$$

$$\sum_{m=0}^{\infty} m \exp\left[-(m+c)^2\right]$$

$$\sim \frac{1}{2} + \sum_{m=0}^{\infty} \frac{(-1)^m}{m!} [\zeta(-2m - 1, c) - c\zeta(-2m, c)]$$

$$- \frac{\sqrt{\pi}}{2} c + \sqrt{\pi} [\pi \sin(2\pi c) - c \cos(2\pi c)] S(\pi^2),$$

(2.72)

$$\sum_{m=0}^{\infty} \exp\left[-a(m+c)^2\right]$$

$$\sim \sum_{m=0}^{\infty} \frac{(-1)^m}{m!} a^m \zeta(-2m, c) + \frac{1}{2}\sqrt{\frac{\pi}{a}} + \sqrt{\frac{\pi}{a}} \cos(2\pi c) S\left(\frac{\pi^2}{a^2}\right).$$

We get the general expression (of Epstein–Hurwitz type with $N = 2$)

$$E_2(s; a_1, a_2; c_1, c_2) \sim \sum_{n_1, n_2 = 0}^{\infty} \left[a_1(n_1 + c_1)^2 + a_2(n_2 + c_2)^2\right]^{-s}$$

$$= \frac{a_2^{-s}}{\Gamma(s)} \sum_{m=0}^{\infty} \frac{(-1)^m \Gamma(s + m)}{m!} \left(\frac{a_1}{a_2}\right)^m$$

$$\cdot \zeta(-2m, c_1) \zeta(2s + 2m, c_2)$$

$$+ \frac{a_2^{-s}}{2} \left(\frac{\pi a_2}{a_1}\right)^{1/2} \frac{\Gamma(s - \frac{1}{2})}{\Gamma(s)} \zeta(2s - 1, c_2)$$

$$\cdot \frac{2\pi^s}{\Gamma(s)} \cos(2\pi c_1) a_1^{-\frac{s}{2} - \frac{1}{4}} a_2^{-\frac{s}{2} + \frac{1}{4}}$$

$$\cdot \sum_{n_1=1}^{\infty} \sum_{n_2=0}^{\infty} n_1^{s-\frac{1}{2}} (n_2 + c_2)^{-s+\frac{1}{2}} K_{s-\frac{1}{2}} \left[2\pi \sqrt{\frac{a_2}{a_1}} n_1 (n_2 + c_2) \right],$$

$$(2.73)$$

K_ν being the modified Bessel function of the second kind. It constitutes the general analytic continuation formula for two-dimensional series. In particular, for $s = 0$,

$$E_2(0; a_1, a_2; c_1, c_2) = \left(c_1 - \frac{1}{2} \right) \left(c_2 - \frac{1}{2} \right), \qquad (2.74)$$

for $s = -1$,

$$E_2(-1; a_1, a_2; c_1, c_2)$$

$$= a_2 \left(\frac{1}{2} - c_1 \right) \zeta(-2, c_2) + a_1 \left(\frac{1}{2} - c_2 \right) \zeta(-2, c_1)$$

$$\cdot \frac{1}{3} \left(c_1 - \frac{1}{2} \right) \left(c_2 - \frac{1}{2} \right) [a_1 c_1 (1 - c_1) + a_2 c_2 (1 - c_2)], \qquad (2.75)$$

and for $s = 2$,

$$E_2(2; a_1, a_2; c_1, c_2)$$

$$\sim \frac{1}{a_2^2} \sum_{m=0}^{\infty} (-1)^m (m + 1) \left(\frac{a_1}{a_2} \right)^m$$

$$\cdot \zeta(-2m, c_1) \zeta(2m + 4, c_2) + \frac{\pi}{4a_2} \frac{1}{\sqrt{a_1 a_2}} \zeta(3, c_2)$$

$$+ \frac{\pi^2 \cos(2\pi c_1)}{a_1 a_2} \sum_{n=0}^{\infty} \left\{ (n + c_2)^{-2} \left[\exp \left(2\pi \sqrt{\frac{a_2}{a_1}} (n + c_2) \right) - 1 \right]^{-2} \right.$$

$$\left. + \left[(n + c_2)^{-2} + \sqrt{\frac{a_1}{a_2}} \frac{(n + c_2)^{-3}}{2\pi} \right] \left[\exp \left(2\pi \sqrt{\frac{a_2}{a_1}} (n + c_2) \right) - 1 \right]^{-1} \right\}.$$

$$(2.76)$$

The general expression for arbitrary N turns out to be

$$E_N(s; a_1, \ldots, a_N; c_1, \ldots, c_N)$$

$$\sim \frac{1}{\Gamma(s)} \sum_{m=0}^{\infty} \frac{(-1)^m}{m!} a_1^m \zeta(-2m, c_1)$$

$$\cdot \Gamma(s + m) E_{N-1}(s + m; a_2, \ldots, a_N; c_2, \ldots, c_N)$$

$$+ \frac{1}{2} \sqrt{\frac{\pi}{a_1}} \frac{\Gamma(s - \frac{1}{2})}{\Gamma(s)} E_{N-1} \left(s - \frac{1}{2}; a_2, \ldots, a_N; c_2, \ldots, c_N \right) + \sqrt{\frac{\pi}{a_1}} \frac{\cos(2\pi c_1)}{\Gamma(s)}$$

$$\cdot \sum_{n_1=1}^{\infty} \sum_{n_2,\ldots,n_N=0}^{\infty} \int_0^{\infty} dt\, t^{s-3/2} \exp\left[-\frac{\pi^2 n_1^2}{a_1 t} - t \sum_{j=2}^{N} a_j (n_j + c_j)^2\right]. \quad (2.77)$$

A word about the notation. In this section we have tried to be consistent with the sign $\sim$ to denote 'asymptotic expansion'. Usually, however, the equality sign is also employed (as for ordinary Taylor series). Normally this does not turn out to be a problem since—at least in the situations to be considered here—the asymptotic (resp. convergent) character of the series on the r.h.s. is not difficult to recognize.

2.4 Expressions for Multi-series on Combinations Involving Arbitrary Constants and Exponents

We shall now make use of the zeta function regularization theorem in order to obtain expressions for the most general multi-series of the type presented in the introduction, which would be impossible to derive by other means (at least with comparable easiness and universality). The same notation which has commonly been used in other references of the author will be employed here, e.g.,

$$M_N^c(s; \vec{a}; \vec{\alpha}; \vec{c}) \equiv M_N^c(s; a_1, \ldots, a_N; \alpha_1, \ldots, \alpha_N; c_1, \ldots, c_N)$$

$$\equiv \sum_{n_1,\ldots,n_N=0}^{\infty} \left[a_1 (n_1 + c_1)^{\alpha_1} + \cdots + a_N (n_N + c_N)^{\alpha_N} + c\right]^{-s},$$

$$(2.78)$$

and for the generalized Epstein-like case:

$$E_N^c(s; \vec{a}; \vec{c}) \equiv M_N^c(s; a_1, \ldots, a_N; 2, \ldots, 2; c_1, \ldots, c_N)$$

$$= \sum_{n_1,\ldots,n_N=0}^{\infty} \left[a_1 (n_1 + c_1)^2 + \cdots + a_N (n_N + c_N)^2 + c\right]^{-s}. \quad (2.79)$$

Consider the case of M_2^c. We need the result of the regularization theorem as applied to the double series

$$S_\alpha(t, s) = \sum_{n=1}^{\infty} \frac{1}{n^{s+1}} \sum_{k=0}^{\infty} \frac{(-t)^k}{k!} n^{\alpha k}, \quad \alpha \in R, \quad (2.80)$$

which converges for $\mathrm{Re}(s) > 0$ large enough. We can write

$$S_\alpha(t, s) = \sum_{n=1}^{\infty} \frac{1}{n^{s+1}} \oint_C \frac{dk}{2\pi i} t^k n^{-\alpha k} \Gamma(k), \quad (2.81)$$

where the contour $\mathcal{C}$ consists of the straight line $\text{Re}(k) = k_0$, with k_0 fixed, $0 < k_0 < 1$, and of the semicircumference at infinity on the left of this line (see Fig. 2). The regularization theorem tells us in this case that [109, 110]

$$S_\alpha(t,s) = \sum_{k=0}^{\infty} \frac{(-t)^k}{k!} \zeta(s+1-\alpha k) + \frac{1}{\alpha}\Gamma\left(-\frac{s}{\alpha}\right)t^{-1/\alpha} - \Delta_\alpha(t,s), \quad \frac{s}{\alpha} \notin N,$$

(2.82)

where $\Delta_\alpha(t,s)$ is the contribution of the curved part K of the contour $\mathcal{C}$:

$$\Delta_\alpha(t,s) \equiv \int_K \frac{dk}{2\pi i}\zeta(s+1+\alpha k)\Gamma(k)t^k.$$

(2.83)

With this, we obtain

$$M_2^c(s; \vec{a}; \vec{\alpha}; \vec{c}) = \frac{a_2^{-s}}{\Gamma(s)} \sum_{m=0}^{\infty} \frac{(-1)^m \Gamma(s+m)}{m!} \left(\frac{a_1}{a_2}\right)^m$$

$$\cdot \zeta(-\alpha_1 m, c_1)M_1^{c/a_2}(s+m; 1; \alpha_2; c_2)$$

$$+ \frac{a_2^{-s}}{\alpha_1}\Gamma\left(\frac{1}{\alpha_1}\right)\left(\frac{a_2}{a_1}\right)^{1/\alpha_1}\frac{\Gamma(s-\frac{1}{\alpha_1})}{\Gamma(s)}M_1^{c/a_2}(s-1/\alpha_1; 1; \alpha_2; c_2)$$

$$+ \frac{a_2^{-s}}{\Gamma(s)}\left(\frac{a_2}{a_1}\right)^{1/\alpha_1}\int_K \frac{da}{2\pi i}\zeta(s+1+\alpha_1 a, c_1)$$

$$\cdot M_1^{c/a_2}(s+a; 1; \alpha_2; c_2)\Gamma(a)\Gamma(s+a),$$

(2.84)

and also

$$M_1^c(s; a_1; \alpha_1; c_1) = \frac{c^{-s}}{\Gamma(s)} \sum_{m=0}^{\infty} \frac{(-1)^m \Gamma(s+m)}{m!}\left(\frac{a_1}{c}\right)^m \zeta(-\alpha_1 m, c_1)$$

$$+ \frac{c^{-s}}{\alpha_1}\Gamma\left(\frac{1}{\alpha_1}\right)\left(\frac{c}{a_1}\right)^{1/\alpha_1}\frac{\Gamma(s-\frac{1}{\alpha_1})}{\Gamma(s)}$$

$$+ \frac{c^{-s}}{\Gamma(s)}\left(\frac{c}{a_1}\right)^{1/\alpha_1}\int_K \frac{da}{2\pi i}\zeta(s+1+\alpha_1 a, c_1)\Gamma(a)\Gamma(s+a).$$

(2.85)

It is not difficult to build, from these two expressions, a recurrence leading to the calculation of M_N^c from the knowledge of M_{N-1}^c, and starting with the formula for M_1^c. At each step, this involves a complex integration over a curved contour at infinity, a term which is in general very small compared with the rest.

As the full calculation is rather involved and lengthy, let us here—for the benefit of the reader—just show in detail the first two steps leading to the formula which corresponds to the case when the c's are zero. We shall accumulate the contributions coming from the series commutation into a single (small) term that we will call simply Δ. So this part will not be really taken care of in the course of the calculation, owing to its smallness (see the relevant numerics in the next chapter). The starting point is again the use of the Mellin transform on the power terms of the series considered

$$M_3(s; \vec{a}; \vec{\alpha}; \vec{c} = \vec{0}) = \sum_{n_1,n_2,n_3=1}^{\infty} \left(a_1 n_1^{\alpha_1} + a_2 n_2^{\alpha_2} + a_3 n_3^{\alpha_3}\right)^{-s}, \qquad (2.86)$$

followed by an expansion on the n_1-terms in the integrand exponent. We thus get

$$\sum_{n_1,n_2,n_3=1}^{\infty} \left(a_1 n_1^{\alpha_1} + a_2 n_2^{\alpha_2} + a_3 n_3^{\alpha_3}\right)^{-s}$$

$$= \frac{1}{a_3^s \Gamma(s)} \left\{ \sum_{n_2,n_3=1}^{\infty} \sum_{k_1=0}^{\infty} (-1)^{k_1} \frac{b_1^{k_1}}{k_1!} \zeta(-\alpha_1 k_1) \right.$$

$$\cdot \int_0^{\infty} dt\, t^{s+k_1-1} \exp\left[-t\left(b_2 n_2^{\alpha_2} + b_3 n_3^{\alpha_3}\right)\right]$$

$$\left. + \frac{\Gamma(1/\alpha_1)}{\alpha_1 b_1^{1/\alpha_1}} \sum_{n_2,n_3=1}^{\infty} \int_0^{\infty} dt\, t^{s-(1/\alpha_1)-1} \exp\left[-t\left(b_2 n_2^{\alpha_2} + b_3 n_3^{\alpha_3}\right)\right] \right\}, \qquad (2.87)$$

where $b_j \equiv a_j/a_3$, $j = 1, 2$. By proceeding again in the same way with the n_2-terms of the exponents, we obtain

$$\sum_{n_1,n_2,n_3=1}^{\infty} \left(a_1 n_1^{\alpha_1} + a_2 n_2^{\alpha_2} + a_3 n_3^{\alpha_3}\right)^{-s}$$

$$= \frac{1}{a_3^s \Gamma(s)} \left\{ \sum_{n_3=1}^{\infty} \sum_{k_1,k_2=0}^{\infty} (-1)^{k_1+k_2} \frac{b_1^{k_1}}{k_1!} \frac{b_2^{k_2}}{k_2!} \zeta(-\alpha_1 k_1)\zeta(-\alpha_2 k_2) \right.$$

$$\cdot \int_0^{\infty} dt\, t^{s+k_1+k_2-1} \exp\left(-t n_3^{\alpha_3}\right)$$

$$+ \frac{\Gamma(1/\alpha_1)}{\alpha_1 b_1^{1/\alpha_1}} \sum_{n_3=1}^{\infty} \sum_{k_2=0}^{\infty} (-1)^{k_2} \frac{b_2^{k_2}}{k_2!} \zeta(-\alpha_2 k_2) \int_0^{\infty} dt\, t^{s+k_2-(1/\alpha_1)-1} \exp\left(-t n_3^{\alpha_3}\right)$$

$$+ \frac{\Gamma(1/\alpha_2)}{\alpha_2 b_2^{1/\alpha_2}} \sum_{n_3=1}^{\infty} \sum_{k_1=0}^{\infty} (-1)^{k_1} \frac{b_1^{k_1}}{k_1!} \zeta(-\alpha_1 k_1) \int_0^{\infty} dt\, t^{s+k_1-(1/\alpha_2)-1} \exp\left(-t n_3^{\alpha_3}\right)$$

$$\left. + \frac{\Gamma(1/\alpha_1)}{\alpha_1 b_1^{1/\alpha_1}} \frac{\Gamma(1/\alpha_2)}{\alpha_2 b_2^{1/\alpha_2}} \sum_{n_3=1}^{\infty} \int_0^{\infty} dt\, t^{s-(1/\alpha_1)-(1/\alpha_2)-1} \exp\left(-t n_3^{\alpha_3}\right) \right\}. \qquad (2.88)$$

Performing now the series commutation and resumming the corresponding zeta functions, we finally get the asymptotic series:

$$M_3(s; \vec{a}; \vec{\alpha}; \vec{c} = \vec{0})$$

$$= \sum_{n_1,n_2,n_3=1}^{\infty} \left(a_1 n_1^{\alpha_1} + a_2 n_2^{\alpha_2} + a_3 n_3^{\alpha_3}\right)^{-s}$$

$$\sim \frac{1}{a_3^s \Gamma(s)} \left\{ \sum_{k_1,k_2=0}^{\infty} (-1)^{k_1+k_2} \frac{b_1^{k_1} b_2^{k_2}}{k_1! \, k_2!} \Gamma(s+k_1+k_2) \right.$$

$$\cdot \zeta(-\alpha_1 k_1)\zeta(-\alpha_2 k_2)\zeta\big(\alpha_3(s+k_1+k_2)\big)$$

$$+ \frac{\Gamma(1/\alpha_1)}{\alpha_1 b_1^{1/\alpha_1}} \sum_{k_2=0}^{\infty} (-1)^{k_2} \frac{b_2^{k_2}}{k_2!} \Gamma(s+k_2-1/\alpha_1)\zeta(-\alpha_2 k_2)\zeta\big(\alpha_3(s+k_2-1/\alpha_1)\big)$$

$$+ \frac{\Gamma(1/\alpha_2)}{\alpha_2 b_2^{1/\alpha_2}} \sum_{k_1=0}^{\infty} (-1)^{k_1} \frac{b_1^{k_1}}{k_1!} \Gamma(s+k_1-1/\alpha_2)\zeta(-\alpha_1 k_1)\zeta\big(\alpha_3(s+k_1-1/\alpha_2)\big)$$

$$\left. + \frac{\Gamma(1/\alpha_1)}{\alpha_1 b_1^{1/\alpha_1}} \frac{\Gamma(1/\alpha_2)}{\alpha_2 b_2^{1/\alpha_2}} \Gamma(s-1/\alpha_1-1/\alpha_2)\zeta\big(\alpha_3(s-1/\alpha_1-1/\alpha_2)\big) \right\} + \Delta.$$

$$(2.89)$$

By repeating this analysis, after N steps we easily obtain the general formula (for the $\vec{c} = \vec{0}$, $c = 0$ case, with $b_j \equiv a_j/a_N$, $j = 1, \ldots, N-1$):

$$M_N(s; \vec{a}; \vec{\alpha}; \vec{c} = \vec{0})$$

$$= \sum_{n_1,\ldots,n_N=1}^{\infty} \left(a_1 n_1^{\alpha_1} + \cdots + a_N n_N^{\alpha_N}\right)^{-s}$$

$$\sim \frac{1}{a_N^s \Gamma(s)} \sum_{p=0}^{N-1} \sum_{C_{N-1,p}} \prod_{r=1}^{p} \frac{b_{i_r}^{-1/\alpha_{i_r}}}{\alpha_{i_r}} \Gamma\left(\frac{1}{\alpha_{i_r}}\right)$$

$$\cdot \sum_{k_{j_1},\ldots,k_{j_{N-p-1}}=0}^{\infty} \Gamma\left(s + \sum_{l=1}^{N-p-1} k_{j_l} - \sum_{r=1}^{p} \frac{1}{\alpha_{i_r}}\right) \prod_{l=1}^{N-p-1} \frac{(-b_{j_l})^{k_{j_l}}}{k_{j_l}!}$$

$$\cdot \zeta(-\alpha_{j_l} k_{j_l})\zeta\left(\alpha_N \left[s + \sum_{l=1}^{N-p-1} k_{j_l} - \sum_{r=1}^{p} \frac{1}{\alpha_{i_r}}\right]\right) + \Delta_{ER},$$

$$(2.90)$$

where $1 \le i_1 < \cdots < i_p \le N-1$, $1 \le j_1 < \cdots < j_{N-p-1} \le N-1$, being $i_1, \ldots, i_p$, $j_1, \ldots, j_{N-p-1}$ a permutation of $1, 2, \ldots, N-1$. The sum on $C_{N-1,p}$ means sum over the $\binom{N-1}{p}$ choices of the indices $i_1, \ldots, i_p$ among the $1, 2, \ldots, N-1$, and the term Δ_{ER} includes all the Δ corrections which appear at each step of the recurrence.

Proceeding in the same way when we are confronted with the general case $\vec{c} \neq \vec{0}$ and $c \neq 0$, the recurrence can be solved explicitly also, the result being ([109], corrected)

$$
M_N^c(s; \vec{a}; \vec{\alpha}; \vec{c}) = \sum_{n_1, \ldots, n_N = 1}^{\infty} \left(a_1 n_1^{\alpha_1} + \cdots + a_N n_N^{\alpha_N} + c \right)^{-s}
$$

$$
\cdot \frac{a_N^{-s}}{\Gamma(s)} \sum_{p=0}^{N-1} \sum_{C_{N-1,p}} \prod_{r=1}^{p} \frac{b_{i_r}^{-1/\alpha_{i_r}}}{\alpha_{i_r}} \Gamma\left(\frac{1}{\alpha_{i_r}}\right)
$$

$$
\cdot \sum_{k_{j_1}, \ldots, k_{j_{N-p-1}} = 0}^{\infty} \Gamma\left(s + \sum_{l=1}^{N-p-1} k_{j_l} - \sum_{r=1}^{p} \frac{1}{\alpha_{i_r}} \right)
$$

$$
\cdot \prod_{l=1}^{N-p-1} \frac{(-b_{j_l})^{k_{j_l}}}{k_{j_l}!} \zeta(-\alpha_{j_l} k_{j_l}, c_{j_l})
$$

$$
\cdot M_1^{c/a_N}\left(\alpha_N\left(s + \sum_{l=1}^{N-p-1} k_{j_l} - \sum_{r=1}^{p} \frac{1}{\alpha_{i_r}} \right); 1; \alpha_N \right) + \Delta_{ER}, \quad (2.91)
$$

(notice that a small mistake in (3.22) and (3.23) of Ref. [109] has been corrected).

Going down to the particular case when the $\alpha_i = 2$ (2.79) things become much more concrete. As mentioned before, then the expression giving our additional corrections to the series commutation reduces to a theta function identity [109], with the result

$$
\sum_{m=0}^{\infty} \exp\left[-a(m+c)^2 \right] \sim \sum_{m=0}^{\infty} \frac{(-1)^m}{m!} a^m \zeta(-2m, c) + \frac{1}{2}\sqrt{\frac{\pi}{a}}
$$

$$
+ \sqrt{\frac{\pi}{a}} \cos(2\pi c) S\left(\frac{\pi^2}{a^2}\right), \quad (2.92)
$$

and this yields the recurrence

$$
E_N^c(s; \vec{a}; \vec{c}) = \frac{1}{\Gamma(s)} \sum_{m=0}^{\infty} \frac{(-1)^m}{m!} a_1^m \zeta(-2m, c_1) \Gamma(s+m)
$$

$$
\cdot E_{N-1}^c(s+m; a_2, \ldots, a_N; c_2, \ldots, c_N)
$$

$$
+ \frac{1}{2}\sqrt{\frac{\pi}{a_1}} \frac{\Gamma(s-1/2)}{\Gamma(s)} E_{N-1}^c(s-1/2; a_2, \ldots, a_N; c_2, \ldots, c_N)
$$

$$
+ \frac{2\pi^s}{\Gamma(s)} \cos(2\pi c_1) a_1^{-s/2-1/4}
$$

$$\cdot \sum_{n_1=1}^{\infty} \sum_{n_2,\dots,n_N=0}^{\infty} n_1^{s-1/2} \left[c + \sum_{j=2}^{N} a_j (n_j + c_j)^2 \right]^{-s/2+1/4}$$

$$\cdot K_{s-1/2} \left(\frac{2\pi n_1}{\sqrt{a_1}} \sqrt{ + \sum_{j=2}^{N} a_j (n_j + c_j)^2 } \right), \qquad (2.93)$$

where K_ν is the modified Bessel function of the second kind. The recurrence starts with

$$E_1^c(s; a_1; c_1) \sim \frac{c^{-s}}{\Gamma(s)} \sum_{m=0}^{\infty} \frac{(-1)^m \Gamma(s+m)}{m!} \left(\frac{a_1}{c} \right)^m \zeta(-2m, c_1)$$

$$+ \frac{c^{1/2-s}}{2} \sqrt{\frac{\pi}{a_1}} \frac{\Gamma(s - \frac{1}{2})}{\Gamma(s)} + \frac{2\pi^s}{\Gamma(s)} \cos(2\pi c_1) a_1^{-s/2-1/4} c^{-s/2+1/4}$$

$$\cdot \sum_{n_1=1}^{\infty} n_1^{s-1/2} K_{s-1/2} \left(2\pi n_1 \sqrt{\frac{c}{a_1}} \right). \qquad (2.94)$$

Then

$$E_2^c(s; a_1, a_2; c_1, c_2)$$

$$\sim \frac{a_2^{-s}}{\Gamma(s)} \sum_{m=0}^{\infty} \frac{(-1)^m \Gamma(s+m)}{m!} \left(\frac{a_1}{a_2} \right)^m \zeta(-2m, c_1)$$

$$\cdot E_1^{c/a_2}(s+m; 1; c_2) + \frac{a_2^{1/2-s}}{2} \sqrt{\frac{\pi}{a_1}} \frac{\Gamma(s - \frac{1}{2})}{\Gamma(s)} E_1^{c/a_2}(s - 1/2; 1; c_2)$$

$$+ \frac{2\pi^s}{\Gamma(s)} \cos(2\pi c_1) a_1^{-s/2-1/4} a_2^{-s/2+1/4}$$

$$\cdot \sum_{n_1=1}^{\infty} \sum_{n_2=0}^{\infty} n_1^{s-1/2} \left[a_2(n_2 + c_2)^2 + c \right]^{-s/2+1/4}$$

$$\cdot K_{s-1/2} \left(\frac{2\pi n_1}{\sqrt{a_1}} \sqrt{a_2(n_2 + c_2)^2 + c} \right), \qquad (2.95)$$

and so on. Expressions for the special case $c = 0$ are given in Ref. [109] (see also the preceding section and equations below).

The very particular case, $a_1 = \cdots = a_N = 1$, $c_1 = \cdots = c_N = 1$ and $\alpha_1 = \cdots = \alpha_N = 2$, simplifies considerably. For $c = 0$, we get

$$E_N(s) = \frac{(-1)^{N-1}}{2^{N-1}} \frac{1}{\Gamma(s)} \sum_{j=0}^{N-1} (-1)^j \binom{N-1}{j} \Gamma(2s - j) \zeta \left(s - \frac{j}{2} \right) + \Delta_{ER}, \qquad (2.96)$$

and, for $c \neq 0$,

$$E_N^c(s) = \frac{(-1)^{N-1}}{2^{N-1}} \frac{1}{\Gamma(s)} \sum_{j=0}^{N-1} (-1)^j \binom{N-1}{j} \Gamma\left(s - \frac{j}{2}\right) E_1^c\left(s - \frac{j}{2}\right) + \Delta_{ER}.$$

(2.97)

The poles of this last function arise from those of $E_1^c(s - j/2)$, which are obtained for the values of s such that $s - j/2 = 1/2, -1/2, -3/2, \ldots$. They are poles of order one at $s = N/2, (N-1)/2, N/2 - 1, \ldots$, except for $s = 0, -1, -2, \ldots$, since then the function is finite (owing to the $\Gamma(s)$ in the denominator). These poles are directly removed by zeta-function regularization.

In the particular case $c_1 = \cdots = c_N = 0$, we have

$$E_N^c(s; a_1, \ldots, a_N)$$

$$\equiv \sum_{n_1,\ldots,n_N=1}^{\infty} \left(a_1 n_1^2 + \cdots + a_N n_N^2 + c^2\right)^{-s}$$

$$= -\frac{1}{2} E_{N-1}^c(s; a_2, \ldots, a_N) + \frac{1}{2}\sqrt{\frac{\pi}{a_1}} \frac{\Gamma(s - 1/2)}{\Gamma(s)} E_{N-1}^c(s - 1/2; a_2, \ldots, a_N)$$

$$+ \frac{\pi^s}{\Gamma(s)} a_1^{-s/2} \sum_{k=0}^{\infty} \frac{a_1^{k/2}}{k!(16\pi)^k} \prod_{j=1}^{k} \left[(2s-1)^2 - (2j-1)^2\right] \sum_{n_1,\ldots,n_N=1}^{\infty} n_1^{s-k-1}$$

$$\cdot \left(a_2 n_2^2 + \cdots + a_N n_N^2 + c^2\right)^{-(s+k)/2}$$

$$\cdot \exp\left[-\frac{2\pi}{\sqrt{a_1}} n_1 \left(a_2 n_2^2 + \cdots + a_N n_N^2 + c^2\right)^{1/2}\right].$$

(2.98)

The recurrence starts from expression

$$E_1^c(s; 1) = -\frac{c^{-2s}}{2} + \frac{\sqrt{\pi}}{2} \frac{\Gamma(s - 1/2)}{\Gamma(s)} c^{-2s+1}$$

$$+ \frac{2\pi^s c^{-s+1/2}}{\Gamma(s)} \sum_{n=1}^{\infty} n^{s-1/2} K_{s-1/2}(2\pi nc).$$

(2.99)

We get, for $c \neq 0$,

$$E_2^c(s) = -\frac{1}{2} E_1^c(s) + \frac{\sqrt{\pi}}{2} \frac{\Gamma(s - \frac{1}{2})}{\Gamma(s)} E_1^c\left(s - \frac{1}{2}\right) + \Delta_{ER},$$

$$E_3^c(s) = \frac{1}{4} E_1^c(s) - \frac{\sqrt{\pi}}{2} \frac{\Gamma(s - \frac{1}{2})}{\Gamma(s)} E_1^c\left(s - \frac{1}{2}\right) + \frac{\pi}{4(s-1)} E_1^c(s - 1) + \Delta_{ER},$$

(2.100)

and similar expressions for $c = 0$. This case can be obtained from the former by analytically continuing in the parameter c. Δ_{ER} is again the well-known term coming from (additional) series commutation. Actually, for numerical evaluations we do not need to consider exponentially small terms in the asymptotic expansions above, which give a very good and quick approximation.

To finish, another couple of particularly useful expressions are

$$\sum_{n_1,n_2=1}^{\infty} \sqrt{\left(\frac{n_1}{a_1}\right)^2 + \left(\frac{n_2}{a_2}\right)^2}$$
$$= \frac{1}{24}\left(\frac{1}{a_1} + \frac{1}{a_2}\right) - \frac{\zeta(3)}{8\pi^2}\left(\frac{a_1}{a_2^2} + \frac{a_2}{a_1^2}\right)$$
$$- \frac{\pi^{3/2}}{2\sqrt{a_1a_2}}\left[\exp\left(-2\pi\frac{a_1}{a_2}\right)\left(1 + \mathcal{O}(10^{-3})\right)\right], \tag{2.101}$$

and (this one obtained after additional regularization)

$$\sum_{n_1,n_2=1}^{\infty} \sqrt{\left(\frac{n_1}{a_1}\right)^2 + \left(\frac{n_2}{a_2}\right)^2 + c^2}$$
$$= \frac{c}{4} - \frac{\pi}{6}a_1a_2c^3 + \left(\frac{1}{4\pi}\sqrt{\frac{c}{a_1}} - \frac{ca_2}{4\pi a_1}\right)\left[\exp(-2\pi ca_1)\left(1 + \mathcal{O}(10^{-3})\right)\right].$$
$$\tag{2.102}$$

In both cases we have assumed (this is, of course, no restriction) that $a_2 \leq a_1$.

Chapter 3
A Treatment of the Non-polynomial Contributions: Application to Calculate Partition Functions of Strings and Membranes

In this chapter we will consider a very interesting way of dealing with the additional term of non-polynomial type that shows up in the series commutation relevant to the zeta-function regularization theorem, as described in the preceding chapters. The asymptoticity of the series is proven for the important cases which are useful in Physics (e.g., sums over non-complete lattices, mainly coming from Neumann and Robin BC) and cannot be dealt with using the otherwise very powerful formulas (as Jacobi's theta function identity) obtained from Poisson's summation in many dimensions. Later, a first physical application to calculate the partition function corresponding to string, membrane and, in general, p-brane theories will be investigated. Such theories are usually termed as fundamental in any attempt at a rigorous description of QED from first principles.

3.1 Dealing with the Non-polynomial Term Δ_{ER}

As we have seen already, zeta-function resummation formulas are at the very heart of the zeta-function regularization procedure. They constitute the key ingredient in the proof of the zeta function regularization theorem. The importance of the final outcome of the theorem—and also, of its extension to multiple, generalized Epstein–Hurwitz series with arbitrary exponents—has been properly stressed before. In particular, it is necessary for practical applications of the theorem to estimate or put a bound to the error one introduces by neglecting the non-polynomial terms which arise in the series commutation process. These terms have proven to be difficult to handle and, due to the fact that they are rather small in many cases, usually they have just been dropped off from the final formulas. To put an end to such an unpleasant situation will be one of the main purposes of the present section.

Another motivation is to introduce a completely different method for the derivation of such additional contributions, which comes about from an elegant approach that has its roots in an elaborate admixture of the Mellin transform technique and the heat kernel method. Our new procedure yields a convenient, closed expression for the zeta function corresponding to elliptic operators in terms of complex integrals

E. Elizalde, *Ten Physical Applications of Spectral Zeta Functions*,
Lecture Notes in Physics 855,
DOI 10.1007/978-3-642-29405-1_3, © Springer-Verlag Berlin Heidelberg 2012

over movable vertical lines in the complex plane. The final result is the same as the one painstakingly obtained *via* the original method of the series commutation—with the great advantage that the proof of asymptoticity of the resulting series turns out to be here easier and also, that the calculation of bounds to the remainder term is somewhat simplified. Moreover, cast in this form the theorem admits a natural generalization to the case where the spectrum of the operator is not known. This method had only been used for the particular (albeit important) case $\operatorname{tr} e^{-tA}$, in Ref. [111]. Specific numerical evaluations and an analytical bound on the elusive additional term which has appeared as a byproduct of the series commutation procedure in previous sections will be obtained here too, together with some convenient integral expressions for this term. As a particular example of the usefulness of the whole procedure, we will also derive expressions for the analytical continuation of the involved zeta-type series which appear in the calculation of the partition functions of strings and membranes. This issue happens to be of some interest in deciding if QCD can (or cannot) in fact be described as a superstring (or supermembrane) theory.

A technically important point concerns the range of the sums in the expressions that provide the analytic continuation of the zeta-function series and which result from the application of the zeta-function regularization theorem, previously considered. In the preceding chapter this range has been formally taken to be infinite: the sums are series and the integration contour is at 'infinite distance' from the origin. Then these expressions can have in general just a formal character: on the one hand the series converges only in specific cases (being in general divergent and quickly oscillating) and, on the other, the contribution of the contour at infinity is usually infinite. This will not be any more our philosophy from now on. Below we shall argue that, under quite general conditions, since the series appearing are asymptotic we should always consider them truncated after the most favorable term (the optimal truncation), and will yield always, in the end, finite sums. The contour C in the zeta regularization theorem will be now a vertical line at a constant finite abscissa c (i.e., $\operatorname{Re} z = c$), together with the two corresponding horizontal segments at infinity. This is the central point of the new ideas here, and must be kept always in mind, from now on, when considering expressions like the ones that appear in the formulation of the zeta-function regularization theorem. In particular, we must distinguish the concept of *remainder* that we will here employ in the most general (and difficult) cases (and corresponds to what is left of the analytical continuation of the function under consideration after writing a finite number of terms for the series, in other words, it is associated with the truncation of the series) from the concept of *additional term* used in the previous chapter (which is a further contribution, to the whole series, and which can be defined and is finite only in some special cases, when the series is a convergent one). Thus the additional term is the limit of the remainder, when it exists. But, in general, only the remainder can be defined and the series whose truncations are the few first terms that we shall consider are asymptotic. Rigorous specifications of this point will be given in the following sections.

3.1.1 Proof of Asymptoticity of the Series

Of the three terms which appear in the series commutation (a basic ingredient for the proof of the zeta regularization theorem)—i.e., the one arising from normal commutation, the ordinary contributions of polynomial type and the additional one which is not polynomial and we have described as a contribution of the contour at infinity (see Chap. 2)—this last one, that we have called Δ_{ER}, was historically the most difficult to tackle. In fact, as described in Chap. 2, it was completely overlooked in the first formulations of the zeta function regularization theorem, what led to several erroneous results. It can be interpreted as the remainder of the series that precedes it (and which will be easily proven here to be asymptotic). In its initial form—as we have seen before—it requires an evaluation of the integrand—which involves the zeta function of the elliptic operator—all the way along a contour of infinite radius. This has certainly sense (as we have seen with concrete examples in detail), but only in some specific cases. We shall here show explicitly how to deal in general with the asymptotic expansion of expressions in which the residual term consists precisely of the third term (the elusive one), by performing a complex integral over a vertical line of constant, finite abscissa.

Formally, the best way to proceed is through the Mellin transform $M[f]$. In order to determine the behavior of $M[f](x + iy)$ for $|y| \to \infty$ the following standard theorems will be useful (they can be found in any standard book on asymptotic series as, for instance, in [84, 99, 112])

Theorem 1 *Let $f \in \mathbb{C}^n(0, \infty)$ and suppose there exists $x_0 \in \mathbb{R}$ such that, for $x > x_0$, $\lim_{t \to \infty}(td/dt)^p(t^x f) = 0$, for $p = 0, 1, 2, \ldots, n$, and that $t^{-1} f_p$ is absolutely integrable, where $f_p(t, x) \equiv (td/dt)^p(t^x f)$. Then $M[f](z) = \mathcal{O}(|y|^{-n})$ for $y \to \pm\infty$.*

Proof Follows immediately by wishful application of Riemann–Lebesgue's lemma. $\qquad\square$

Theorem 2 *Let $f \in \mathbb{C}^n(0, \infty)$ with the asymptotic behavior for $t \to 0^+$: $f(t) \sim \sum_{m=0}^{\infty} p_m t^{a_m}$, with $\mathrm{Re}\, a_m$ monotonically increasing towards infinity. Suppose also that asymptotic expansions for $t \to 0^+$ for the successive derivatives of f are obtained by taking the derivative, term by term, of the above expansion for f. Suppose also that $\lim_{t \to \infty} f_p(t, x) = 0$, for $p = 0, 1, 2, \ldots, n$, and that $t^{-1} f_p$ is absolutely integrable for $x > -\mathrm{Re}\, a_0$. Then it follows that $M[f](z) = \mathcal{O}(|y|^{-n})$ for $y \to \pm\infty$ and for any x, that is, in the whole domain of the function, after its continuation.*

Proof Let $\rho \in \mathbb{R}$ be large enough and μ such that $\mathrm{Re}\, a_{\mu-1} < \rho \leq \mathrm{Re}\, a_\mu$, where for a positive integer δ satisfying $\mathrm{Re}\, a_0 + \delta > \mathrm{Re}\, a_\mu$, we define $\sigma_\rho(t) \equiv \exp(-t^\delta) \times \sum_{m=0}^{\mu-1} p_m t^{a_m}$, and $\hat{f} \equiv f - \sigma_\rho$. Applying Theorem 1 to $\hat{f}$ and noticing that $M[\sigma]_\rho(z) = \sum_{m=0}^{\mu-1}(p_m/\delta)\Gamma((z+a_m)/\delta)$ has an exponential behavior for $|y| \to \infty$, from $M[f] = M\hat{f} + M[\sigma]_\rho$ (which gives the continuation to $-\mathrm{Re}\, a_\mu < \mathrm{Re}\, z < \beta$) and the fact that ρ is arbitrarily large we prove the theorem. $\qquad\square$

For $x_0, \theta_0 \in \mathbb{R}$, let us define that a function h belongs to K, $h \in K(x_0, \theta_0)$ iff for any $x > x_0$ and $\epsilon > 0$ one has $M[h](x + iy) = \mathcal{O}[\exp(-(\theta_0 - \epsilon)|y|)]$, for $|y| \to \infty$. Also, denote the sector $S(\theta_0) = \{t \neq 0; |\arg t| < \theta_0\}$.

Theorem 3 *Suppose that in the sector $S(\theta_0)$: (i) h is analytic; (ii) $h = \mathcal{O}(t^\alpha)$, $t \to 0^+$; and (iii) $h = \exp(-at^\nu) \sum_{m=0}^{\infty} \sum_{n=0}^{N(m)} c_{mn} (\ln t)^n t^{-r_m}$, $t \to \infty$, with $\mathrm{Re}\, a \geq 0$, $\nu > 0$ and $\mathrm{Re}\, r_m$ monotonically increasing towards infinity. Then: (i) if $a = 0$, $h \in K(-\mathrm{Re}\, \alpha, \theta_0)$; (ii) if $\mathrm{Re}\, a > 0$, $h \in K(-\mathrm{Re}\, \alpha, \theta)$, where $\theta = \min(\theta_0, (\pi - 2|\arg a|)/(2\nu))$.*

Proof We only do the first case since the second is analogous. Observe that $M[h](z)$ is analytical for $\mathrm{Re}\, z > -\mathrm{Re}\, \alpha$. Define $\theta' = \theta - \epsilon$, θ as given in the theorem and $\epsilon > 0$; deforming the circuit by an angle $\pm \theta'$ and making the change of variable $t = re^{\pm i\theta'}$ (r, real, will be the new variable of integration) in the Mellin transform integral, one easily shows that $M[h](z) = \mathcal{O}[\exp(-\theta'|y|)]$, for $|y| \to \infty$, $x > \mathrm{Re} -\alpha$.

With a similar strategy as that used in Theorem 2, with convenient choices of the function σ_ρ, one can prove results analogous to Theorem 3 (so-called *theorems of exponential decrease*) valid for a range of x which moves towards the left. The only proviso is that h have a convenient asymptotic expansion around 0.

Also to be noticed is the fact that in order to apply these theorems, with the aim of displacing the contour of integration, one must make sure that the conclusions of Theorems 1–3 are valid uniformly on any segment of the x variable along which one wants to displace the integration contour. If we can perform the translation from $x = c$ to $x = c'$, $c' < c$, and we check that $\zeta_A(s + c' + iy)M[f](c' + iy)$ is absolutely integrable, then

$$F(s,t) = \sum_{c' < \mathrm{Re}\, z < c} \mathrm{Res}\left[\zeta_A(s + z)t^{-z}M[f](z)\right]$$

$$+ \frac{t^{-c'}}{2\pi i} \int_{\Re z = c'} dz\, \zeta_A(s + z)t^{-iy}M[f](z). \qquad (3.1)$$

This expression contains the key to the asymptoticity of the series. It shows in fact that the series is asymptotic, since it is immediate that the last term is of small order with respect to the order of the terms of the finite series.   □

To finish this subsection, let us consider a different example which cannot be resolved by direct application of Theorem 3 above (since it corresponds to $a \neq 0$ and $\mathrm{Re}\, a = 0$).

Example 1 Let

$$F(s,t) = \sum_{n=1}^{\infty} \frac{J_\mu(nt)}{n^s}, \qquad (3.2)$$

that is, $\lambda_n = n$, $f = J_\mu$, $\alpha = -\mu$, $\beta = 1/2$. We have

$$M[J]_\mu(z) = \frac{2^{z-1}\Gamma((z+\mu)/2)}{\Gamma((\mu-z+2)/2)}, \tag{3.3}$$

with $s_0 = 1$, $\operatorname{Re} s > 1/2$ (in this case the domain of s can be enlarged), and $c > 1 - \operatorname{Re} s < 1/2$. $M[J]_\mu$ has poles for $z = -\mu + 2n$, $n \in \mathbb{N}$, with residues $(-1)^n 2^{-\mu-2n}/[n!(\mu+n)!]$, while $\zeta(s+z)$ has a pole at $z = -s + 1$. From (3.3) and from the behavior of the gamma function we know that in any closed interval of $x = \operatorname{Re} z$ we get the behavior

$$M[J]_\mu(z) = \mathcal{O}(|y|^{x-1/2}), \quad |y| \to \infty, \tag{3.4}$$

uniformly in x. We have $\operatorname{Re} s = 1/2 + \epsilon_1$, $\epsilon_1 > 0$, and choose $c = 1/2 - \epsilon_2$, with $0 < \epsilon_2 < \epsilon_1$, and any δ, $0 < \delta < \min\{1/2, \epsilon_1 - \epsilon_2\}$.

Consider now the following, segmentwise, displacement

$$G(z) = M[J]_\mu(z)\zeta(s+z). \tag{3.5}$$

Then it turns out that:

1. For $1/2 + \delta - \epsilon_1 \le x \le 1/2 - \epsilon_2$, $|M[J]_\mu(z)| = \mathcal{O}(|y|^{-1/2-\epsilon_2})$ and $|\zeta(s+z)| = \mathcal{O}(1)$, therefore

$$\left|G(z)\right| = \mathcal{O}(|y|^{-1/2-\epsilon_2}). \tag{3.6}$$

2. For $1/2 - \epsilon_1 \le x \le 1/2 + \delta - \epsilon_1$, $|M[J]_\mu(z)| = \mathcal{O}(|y|^{-1/2-\epsilon_1})$ and $|\zeta(s+z)| = \mathcal{O}(\ln|y|)$, therefore

$$\left|G(z)\right| = \mathcal{O}(|y|^{-1/2-\epsilon_1}\ln|y|). \tag{3.7}$$

3. For $-1/2 + \delta - \epsilon_1 \le x \le 1/2 - \epsilon_1$, $|M[J]_\mu(z)| = \mathcal{O}(|y|^{-3/2+\delta-\epsilon_1})$ and $|\zeta(s+z)| = \mathcal{O}(|y|^{1-\delta}\ln|y|)$, therefore

$$\left|G(z)\right| = \mathcal{O}(|y|^{-1/2-\epsilon_1}\ln|y|). \tag{3.8}$$

4. For $-1/2 - \epsilon_1 \le x \le -1/2 + \delta - \epsilon_1$, $|M[J]_\mu(z)| = \mathcal{O}(|y|^{-3/2+\delta-\epsilon_1})$ and $|\zeta(s+z)| = \mathcal{O}(|y|^{1/2})$, therefore

$$\left|G(z)\right| = \mathcal{O}(|y|^{-1+\delta-\epsilon_1}). \tag{3.9}$$

5. Finally, for $-K - 1/2 - \epsilon_1 \le x \le -1/2 - \epsilon_1$, $|M[J]_\mu(z)| = \mathcal{O}(|y|^{-x-1})$ and $|\zeta(s+z)| = \mathcal{O}(|y|^{-x-\epsilon_1})$, therefore

$$\left|G(z)\right| = \mathcal{O}(|y|^{-1-\epsilon_1}), \tag{3.10}$$

where K is arbitrarily large.

Example 2 Take $\lambda_n = n$ and add a degeneration $g(n)$ (being g an analytical function with good behavior in order to fulfill the conditions), i.e.

$$\zeta_g(z) \equiv \sum_{n=1}^{\infty} \frac{g(n)}{n^z} = -\frac{i}{2\pi} \int_{\mathcal{C}} \ln[\sin(\pi t)] \frac{d}{dt} (t^{-z}g) \, dt, \tag{3.11}$$

being $\mathcal{C}$ the contour of the sector of the complex plane defined by $|\arg z| \leq \theta_0$ and $|z| \geq \epsilon, 0 < \epsilon < 1$. Decomposing $\mathcal{C}$ into its upper and lower parts, $\mathcal{C}^{\pm}$ (those are in the present case the paths mentioned above), and assuming g to be analytical on the sector, we obtain

$$\zeta_g(z) = -\int_{\epsilon}^{\infty} d\rho \, \rho \frac{d}{d\rho} (\rho^{-z}g) - \frac{1}{2}\epsilon^{-z}g(\epsilon)$$

$$- \frac{i}{2\pi} \sum_{\pm} \int_{\mathcal{C}^{\pm}} dt \, \ln(1 - e^{\pm 2\pi it}) \frac{d}{dt} (t^{-z}g). \tag{3.12}$$

In general, the second and third terms on the r.h.s. are integer functions of z, and only the first one needs to be continued to the left in $\mathrm{Re}\, z$. This can be done without difficulty in many cases by exploiting the knowledge of the asymptotic behavior of g at $+\infty$ [10, 11, 14].

Another interesting example is the following, which has its roots in the Hardy–Ramanujan formula.

Example 3 Consider the spectral series

$$F(t) \equiv \sum_{n=1}^{\infty} \ln(1 - e^{-tn}). \tag{3.13}$$

In this case we have $s = 0$ and

$$f(t) = \ln(1 - e^{-t}). \tag{3.14}$$

The Mellin transform is known

$$M[f](z) = -\frac{1}{2\pi i} \int_{\mathrm{Re}\, z = c} dz \, \Gamma(z)\zeta(1+z)t^{-z}, \tag{3.15}$$

with $c > 1$ and, as a consequence,

$$F(t) = -\frac{1}{2\pi i} \int_{\mathrm{Re}\, z = c} dz \, \Gamma(z)\zeta(1+z)\zeta(z)t^{-z}, \tag{3.16}$$

with $c > 1$. Here the corresponding asymptotic series $F^{(1)}(t)$ reduces to one contribution, involving a logarithmic term and a non vanishing remainder. This follows easily from an identity which is well known in the theory (due to Hardy and Ramanujan) of the elliptic modular functions.

$$\sum_{n=1}^{\infty} \ln(1 - e^{-tn}) = -\frac{\pi^2}{6t} - \frac{1}{2}\ln\left(\frac{t}{2\pi}\right) + \frac{t}{24} + \sum_{n=1}^{\infty} \ln(1 - e^{-4\pi^2 n/t}). \quad (3.17)$$

For the sake of completeness, a simple proof of this identity is presented below.

Proof of the Hardy–Ramanujan formula The proof of the Hardy–Ramanujan formula is very instructive, because it makes use of the Mellin representation, namely

$$F(t) \equiv \sum_{n=1}^{\infty} \ln(1 - e^{-tn}) = -\frac{1}{2\pi i} \int_{\mathrm{Re}\, z=c} dz\, \Gamma(z)\zeta(1+z)\zeta(z)t^{-z}, \quad (3.18)$$

with $c > 1$. Shifting the line of integration from $\mathrm{Re}\, z = c > 1$ to $\mathrm{Re}\, z = c'$, $-1 < c' < 0$, noting that the integrand has a pole of first-order at $z = 1$ and one of second-order at $z = 0$, one arrives at

$$F(t) = -\frac{\pi^2}{6t} - \frac{1}{2}\ln\left(\frac{t}{2\pi}\right) - \frac{1}{2\pi i} \int_{\mathrm{Re}\, z=c'} dz\, \Gamma(z)\zeta(1+z)\zeta(z)t^{-z}. \quad (3.19)$$

Performing the change of variable in the complex integral: $z = -s$, and using the functional equations for the functions $\Gamma(s)$ and $\zeta(s)$ (see Chap. 1), we get

$$F(t) = -\frac{\pi^2}{6t} - \frac{1}{2}\ln\left(\frac{t}{2\pi}\right) - \frac{1}{2\pi i} \int_{\mathrm{Re}\, z=c''} dz\, \Gamma(z)\zeta(1+z)\zeta(z)\left(\frac{4\pi^2}{t}\right)^{-z}, \quad (3.20)$$

where $0 < c'' < 1$. The identity (3.17) is obtained shifting the line of integration to $\mathrm{Re}\, z = c''' > 1$, and taking the first-order pole at $z = 1$ into account.   □

3.1.2 The Remainder Term and the Poisson Resummation Formula

We would like first to examine a particular but important case, as an illustration of the difficulties associated with the determination of the additional term. Such case will be considered also in the next section but under a somewhat different perspective.

Let us consider the selfadjoint operator $A = H^\beta$, where H is the Dirichlet Laplacian on $[0, 1]$ with eigenvalues $\lambda_n = n^2$. Notice that A is *not* a differential operator, unless β is a positive integer. Putting $2\beta = \alpha$, we can write

$$\zeta_A(z) = \sum_{n=1}^{\infty} n^{-\alpha z} = \zeta(\alpha z). \quad (3.21)$$

The related heat kernel trace reads ($t > 0$)

$$\text{Tr}\, e^{-tA} = \sum_n e^{-n^\alpha t} \equiv S_\alpha(t). \tag{3.22}$$

Making use of the Mellin representation one obtains

$$S_\alpha(t) = \frac{1}{2\pi i} \int_C dz\, t^{-z} \zeta(\alpha z) \Gamma(z) = \frac{\Gamma(\frac{1}{\alpha})}{\alpha t^{\frac{1}{\alpha}}} + \sum_{k=0}^{\infty} \frac{(-t)^k}{k!} \zeta(-\alpha k) + \Delta_\alpha(t)$$

$$= \frac{\Gamma(\frac{1}{\alpha})}{\alpha t^{\frac{1}{\alpha}}} - \frac{1}{2} + \sum_{k=1}^{\infty} \frac{(-t)^k}{k!} \zeta(-\alpha k) + \Delta_\alpha(t), \tag{3.23}$$

where the power series in t will converge for $|t| < b$ (for some b, abscissa of convergence). This last expression makes sense, in general, only when $b > 0$. Usually (i.e., unless $\alpha \leq 1$ or $\alpha \in 2\mathbb{N}$) one finds that the series in (3.23) is divergent for any $t \neq 0$ (thus, the abscissa of convergence is $b = 0$). We shall here restrict ourselves to the mentioned particular values of α. The quantity $S_\alpha(t)$ may be evaluated by making use of the Poisson formula, which states that for $f(x) = f(-x)$ and $f(x) \in L_1$, the following equation holds:

$$\sum_{n=1}^{\infty} f(n) = -\frac{1}{2} f(0) + \int_0^\infty dx\, f(x) + 2 \sum_{n=1}^{\infty} \int_0^\infty dx\, f(x) \cos(2\pi nx). \tag{3.24}$$

Let us consider the function

$$f(x) = e^{-t|x|^\alpha}. \tag{3.25}$$

An elementary computation permits us to write

$$S_\alpha(t) = -\frac{1}{2} + \frac{\Gamma(\frac{1}{\alpha})}{\alpha t^{\frac{1}{\alpha}}} + 2 \sum_{n=1}^{\infty} \int_0^\infty dx\, e^{-|x^\alpha|t} \cos(2\pi nx) \tag{3.26}$$

and by comparing (3.23) and (3.26), we get

$$\Delta_\alpha(t) = 2 \sum_{n=1}^{\infty} \int_0^\infty dx\, e^{-|x^\alpha|t} \cos(2\pi nx) - \sum_{k=1}^{\infty} \frac{(-t)^k}{k!} \zeta(-\alpha k). \tag{3.27}$$

The above expression can be checked immediately. In fact, for $\alpha = 2$, i.e. $\beta = 1$, one has $\zeta(-2k) = 0$ and the additional term is nonvanishing

$$\Delta_2(t) = 2 \sum_{n=1}^{\infty} \int_0^\infty dx\, e^{-x^2 t} \cos(2\pi nx) = \sqrt{\frac{\pi}{t}} \sum_{n=1}^{\infty} e^{-\frac{\pi^2 n^2}{t}}. \tag{3.28}$$

As a consequence, (3.23) becomes the well known Jacobi's theta function identity (see Chap. 2).

Again, for $\alpha = 1$, one can perform exactly the elementary integrations involved, and the result is

$$\sum_{n=1}^{\infty} \frac{2t}{t^2 + 4\pi^2 n^2} = -\sum_{r=1}^{\infty} \frac{t^{2r-1}}{(2r-1)!} \zeta(1-2r). \tag{3.29}$$

The sum on the left hand-side can be done with elementary methods and we end up with

$$\frac{1}{2} \coth(t/2) - \frac{1}{t} = -\sum_{r=1}^{\infty} \frac{t^{2r-1}}{(2r-1)!} \zeta(1-2r) \tag{3.30}$$

which is a well known but certainly non trivial identity.

All this goes through for $\alpha = 2, 4, 6, \ldots$. A connection with the theory of Brownian processes may be established at that point [113]. The instabilities that are known to appear outside the range $0 < \alpha < 1$ and outside these particular values of α can be traced back in our procedure to the fact that both the full series in ζ's and the additional term on the contour at infinity diverge. Then, both are useless, in practice, and we must come back to the new method as stated before. This will be illustrated in the following sections with specific examples.

3.2 Numerical Estimates of the Remainder

In view of the above, we shall here restrict ourselves to $\alpha = 2, 4, 6, \ldots$, so that the series in (3.23) vanishes and, provided that we have already picked up the first two terms on the r.h.s. (coming from the poles of $\zeta(\alpha z)\Gamma(z)$) the remainder does not change as we shift $\mathrm{Re}\, c'$ in (3.1) to the left. In particular, we may say that the additional term is precisely this value of the remainder.

Contrary to the additional term (which was the contribution of the semicircumference 'at infinity' [10, 11, 14], difficult to handle in general), the remainder is computed as the contribution at 'finite distance' (an integral along $\mathrm{Re}\, z = c$). The additional term can be approached by the contribution of a semicircumference of finite radius $|z| = R$, for R large enough. Let us consider it in this simpler form (here t is set equal to 1)

$$\Delta_\alpha \equiv \frac{1}{2\pi i} \int_K dz\, \Gamma(z)\zeta(\alpha z), \tag{3.31}$$

where K is the semicircumference of finite radius R, $z = Re^{i\theta}$, with θ going from $\theta = \pi/2$ to $\theta = 3\pi/2$. To begin with, this expression is not defined at $\theta = \pi$ and one must start the calculation by using the well known reflection formulas for Γ and ζ. This yields

$$\Delta_\alpha = i \int_K \frac{dz}{z(2\pi)^{1-\alpha z}} \frac{\sin(\pi \alpha z/2)}{\sin(\pi z)} \frac{\Gamma(1-\alpha z)}{\Gamma(-z)} \zeta(1-\alpha z). \tag{3.32}$$

Such quantity is, in principle, well defined but, on the other hand, rather bad behaved. In modulus, the integrand clearly diverges for $|z| \to \infty$ but, due to the quickly oscillating sinus factors the final result turns out to be finite. Actually, it has been proven in Refs. [51, 109], that it has the very specific value

$$\Delta_2 = -\sqrt{\pi}\, S(\pi^2), \quad S(t) \equiv \sum_{n=1}^{\infty} e^{-\pi^2 t}, \tag{3.33}$$

for $\alpha = 2$, and that it can be made < 1 for any value of $\alpha \in 2\mathbb{N}$ and reasonably high values of R. A possible way to handle expression (3.32) is to employ the integral equation

$$\Gamma(z)\zeta(\beta z) = \int_0^{\infty} S_\beta(t) t^{z-1} dt, \quad \mathrm{Re}\, z > 0, \quad S_\beta(t) \equiv \sum_{n=1}^{\infty} e^{-n^\beta t}, \tag{3.34}$$

what yields

$$\Delta_\alpha = -i\alpha \int_K dz \, \frac{\sin(\pi\alpha z/2)}{\sin(\pi z)} \frac{\Gamma(-\alpha z)}{\Gamma(-z)\Gamma(\frac{1}{\alpha}-z)} \int_0^{\infty} S_\alpha\big((2\pi)^\alpha t\big) t^{\frac{1}{\alpha}-z-1} dt. \tag{3.35}$$

Using now Stirling's formula and a simple approximation for the sinus fraction we get

$$\Delta_\alpha = -\frac{i}{\sqrt{2\pi\alpha}} \int_K dz \, \varphi_\alpha(z) \int_0^{\infty} S_\alpha\big((2\pi/\alpha)^\alpha t\big) t^{\frac{1}{\alpha}-z-1} dt, \tag{3.36}$$

being

$$\varphi_\alpha(z) \equiv \exp\left\{ \left[(2-\alpha)z + \left(\frac{1}{2}-\frac{1}{\alpha}\right)\right] \ln z + (\alpha-2)z + \left(\frac{\alpha}{2}-1\right)\pi \,|\mathrm{Im}\, z| \right.$$
$$\left. + i\,\mathrm{sgn}(\mathrm{Im}\, z)\left(\frac{\alpha}{2}-1\right)\pi\,|\mathrm{Re}\, z| \right\}. \tag{3.37}$$

We see that, in fact, for $\alpha < 1$, $\varphi_\alpha(z) \to 0$, when $|z| \to \infty$, while for $\alpha = 2$ it is $\varphi_2(z) \equiv 1$ and $\Delta_2 = -\sqrt{\pi}\, S_2(\pi^2)$, as anticipated before (this is nothing but Jacobi's theta function identity, see Chap. 2). On the other hand, if we substitute in (3.36) its mean value μ for the function φ_α, we obtain

$$\Delta_\alpha = -\mu \sqrt{\frac{2\pi}{\alpha}} S_\alpha\left(\left(\frac{2\pi}{\alpha}\right)^\alpha\right). \tag{3.38}$$

An analytical but approximate evaluation carried out by the author for the first particular cases $\alpha = 2, 4, 6, 8$, suggests that the behavior of the mean value μ in terms of α can be bounded from above by an expression of the kind

$$|\mu| \le \frac{e-2}{2e} \sqrt{\frac{2\pi}{\alpha}} \tag{3.39}$$

for $\alpha \in 2\mathbb{N}$ large. This gives for the additional term

$$|\Delta_\alpha| \le \frac{e-2}{2e} S_\alpha \left(\left(\frac{2\pi}{\alpha} \right)^\alpha \right), \tag{3.40}$$

a bound that is certainly convergent for $\alpha \to \infty$.

We have checked this analytical bound with a numerical, direct evaluation of the integral (3.32) written in the following equivalent form (obtained with straightforward manipulations)

$$\Delta_\alpha = -\int_{-\pi/2}^{\pi/2} \frac{d\theta}{(2\pi)^{z+1}} \frac{\sin(\pi z/2)}{\sin(\pi z/\alpha)} \frac{\Gamma(z+1)}{\Gamma(z/\alpha)} \zeta(z+1), \quad z = Re^{i\theta}, \ R \gg 1. \tag{3.41}$$

In particular, the case $\alpha \to \infty$ can be easily handled

$$\Delta_\infty = -\frac{1}{\pi} \int_{-\pi/2}^{\pi/2} \frac{d\theta}{(2\pi)^{z+1}} \sin(\pi z/2)\Gamma(z+1)\zeta(z+1), \quad z = Re^{i\theta}, \ R \gg 1. \tag{3.42}$$

As discussed before, it is quite difficult to evaluate such integrals for R big. Any numerical procedure breaks down for R large enough. But, by using this technique we never need, in practice, to go to very large Rs in the evaluation of the additional term—at the expense of just calculating a few first terms of an asymptotic expansion on the zeta function of the differential operator one is dealing with (the asymptotic series must be optimally truncated in the usual way). Putting it in the particular situation which concerns us here, we need only obtain numerical values for the expressions (3.41) and (3.42) when R is just of the order of 20 or 30, what is already involved enough since the usual asymptotic series that appear in practice yield their best result for a number of terms of this (and very often lesser) order.

There seems to be no definite regular behavior in the numerical values of the expressions above, though all of them, in the wide range $8 \le R \le 30$, lie in the quite narrow interval $0.02 \le |\Delta| \le 0.07$. This is in very good agreement with the results obtained from the analytical approximation (3.40), which are $\Delta_4 \simeq 0.04$, $\Delta_6 \simeq 0.07$ and $\Delta_\infty \simeq 0.13$. Only the tendency of this formula for large α deviates from the numerical result by a factor of 10—namely the numerical results are about ten times lower than the upper bound values given by the formula (3.40). But this is not strange, since for big α's expression (3.40) is rather bad (S_α is then a very slowly convergent series), while, on the contrary, the limiting expression for Δ_∞ (3.42) remains in very good shape.

Summing up, (3.41), (3.42) and (3.40) are the best we could obtain, using an analytical approximation, for the treatment of the additional term which appears either in the series commutation procedure or in the Mellin transformation method. For the numerical treatment, the calculation must be performed *not* at $R = \infty$ but at finite, increasing values of R, and in this case (3.41) and (3.42) can be handled through standard numerical integration procedures.

3.3 Application: Summation of the String Partition Function for Different Ranges of the Temperature

The problem of deciding if QCD is actually (some limit of) a theory of extended objects (strings, membranes or p-branes in general) is almost forty years old and goes back to 't Hooft. J. Polchinski tried to answer this question both for the Nambu–Goto and for the rigid strings by calculating their partition functions and seeing if they actually match with the one corresponding to QCD for different ranges of temperature [114, 115]. We shall not go into the particularities of the procedure [114, 115]—which has been described in [116] in some detail. Rather, we shall here concentrate in the specific aspects of the problem that have to do with the zeta-function regularization procedure, only.

The first two terms in the loop expansion

$$S_{eff} = S_0 + S_1 + \cdots \tag{3.43}$$

of the effective action corresponding to the rigid string

$$S = \frac{1}{2\alpha_0} \int d^2\sigma \left[\rho^{-1} \partial^2 X^\mu \partial^2 X_\mu + \lambda^{ab} \left(\partial_a X^\mu \partial_b X_\mu - \rho \delta_{ab} \right) \right] + \mu_0 \int d^2\sigma \, \rho \tag{3.44}$$

being α_0 the dimensionless, asymptotically free coupling, ρ the intrinsic metric, μ_0 the explicit string tension (important at low energy) and λ^{ab} the Lagrange multipliers—are given, in the world sheet $0 \leq \sigma^1 \leq L$ and $0 \leq \sigma^2 \leq \beta t$ (an annulus of modulus t), by

$$S_0 = \frac{L\beta t}{2\alpha_0} \left[\lambda^{11} + \lambda^{22} t^{-2} + \rho \left(2\alpha_0 \mu_0 - \lambda^{aa} \right) \right] \tag{3.45}$$

at the tree level, and by

$$S_1 = \frac{d-2}{2} \ln \det \left(\partial^4 - \rho \lambda^{ab} \partial_a \partial_b \right)$$

$$= \frac{d-2}{2} L \sum_{n=-\infty}^{\infty} \int_{-\infty}^{+\infty} \frac{dk}{2\pi} \ln \left[\left(k^2 + \frac{4\pi^2 n^2}{\beta^2 t^2} \right)^2 + \rho \left(\lambda^{11} k^2 + \frac{4\pi^2 n^2}{\beta^2 t^2 \lambda^{22}} \right) \right] \tag{3.46}$$

at one-loop order. This is quite a non-trivial calculation, which has been performed in Ref. [114, 115] only in the strict limits $T \to 0$ and $T \to \infty$ around some extremizing configuration, the parameters being ρ, λ^{11}, λ^{22} and t. As a first step of the zeta function method we write

$$S_1 = -(d-2) \frac{d}{ds} \zeta_A(s/2) \bigg|_{s=0},$$

$$\zeta_A(s/2) = L \sum_{n=-\infty}^{\infty} \int_{-\infty}^{+\infty} \frac{dk}{2\pi} \left(k^2 + y_+^2 \right)^{-s/2} \left(k^2 + y_-^2 \right)^{-s/2}, \tag{3.47}$$

where

$$y_\pm = \frac{a}{t}\left[n^2 + \frac{\rho t^2 \lambda^{11}}{2a^2} \pm \sqrt{\rho}\frac{t}{a}\left((\lambda^{11} - \lambda^{22})n^2 + \frac{\rho t^2 \lambda^{11^2}}{4a^2}\right)^{1/2}\right]^{1/2}, \quad a \equiv \frac{2\pi}{\beta}.$$

(3.48)

For convenience, one may consider two different approximations of overlapping validity, one for low temperature, $\beta^{-2} \ll \mu_0$, and the other for high temperature, $\beta^{-2} \gg \alpha_0\mu_0$. Both these approximations (overlapping included) are obtainable from the expression above. We choose to write it in the form

$$\zeta_A(s/2) = \frac{L}{2\sqrt{\pi}}\frac{\Gamma(s-1/2)}{\Gamma(s)}\sum_{n=-\infty}^{\infty}\frac{y_-}{(y_+y_-)^s}F(s/2, 1/2; s; 1-\eta), \quad \eta \equiv \frac{y_-^2}{y_+^2}.$$

(3.49)

This is a strict equality. For high temperature, the ordinary expansion of the confluent hypergeometric function F is in order

$$F(s/2, 1/2; s; 1-\eta) = \sum_{k=0}^{\infty}\frac{(s/2)_k(1/2)_k}{k!(s)_k}(1-\eta)^k \longrightarrow \frac{1}{2}(1+\eta^{-1/2}), \quad s \to 0,$$

(3.50)

since $y_\pm$ can be written as

$$y_\pm \simeq \frac{a}{t}\left[(n\pm b)^2 + c^2\right]^{1/2}, \quad b = \frac{\beta t}{4\pi}\sqrt{\rho(\lambda^{11} - \lambda^{22})}, \quad c \simeq \frac{\alpha_0}{4\pi}.$$

(3.51)

After the appropriate analytic continuation, the derivative of the zeta function yields

$$\frac{d}{ds}\zeta_A(s/2)|_{s=0}$$

$$= -\frac{L}{4}\sqrt{b^2 + c^2} + \frac{2\pi L}{\beta t}\left\{b^2 + \frac{1}{6} - \left[\frac{1}{2}\ln(-b^2) + \psi(-1/2) + \gamma\right]c^2\right\}. \quad (3.52)$$

In order to obtain this result, which comes from elementary Hurwitz zeta functions, $\zeta(\pm 1, b)$, we have used in (3.51) the binomial expansion, what is completely consistent with the approximation (notice the extra terms coming from the contribution of the pole of (3.51) to the derivative of $\zeta_A(s/2)$).

The low-temperature case is more involved and now the methods developed in the preceding section will prove to be useful. The term $n = 0$ must be treated separately. It gives [116]

$$\zeta_A^{(n=0)}(s/2) = \frac{L}{2\pi}\frac{\Gamma((1-s)/2)\Gamma(s-1/2)}{\Gamma(s/2)}(\lambda^{11}\rho)^{1/2-s}, \quad \frac{1}{2} < \mathrm{Re}(s) < 1.$$

(3.53)

This is again a closed result, which yields

$$\frac{d}{ds}\zeta_A^{n=0}(s/2)\bigg|_{s=0} = -\frac{L}{2}\sqrt{\lambda\rho} \quad (3.54)$$

and

$$S_1^{n=0} = (d-2)\frac{L}{2}\sqrt{\lambda\rho}. \tag{3.55}$$

Such contribution must be added to the one coming from the remaining terms ((3.49) above)

$$\zeta_A'(s/2) = \frac{L}{\sqrt{\pi}}\frac{\Gamma(s-1/2)}{\Gamma(s)}\sum_{n=1}^{\infty}\frac{y_-}{(y_++y_-)^s}F(s/2,1/2;s;1-\eta), \quad \mathrm{Re}(s) > 1. \tag{3.56}$$

(The prime is no derivative, it just means that the term $n=0$ is absent here.) Within this approximation, and working around the classical, $T=0$ solution: $\lambda^{11}=\lambda^{22}=\alpha_0\mu_0$, $\rho=t^{-2}=1$, we obtain

$$\zeta_A'(s/2) = \frac{L}{\sqrt{\pi}}\frac{\Gamma(s-1/2)}{\Gamma(s)}\left(\frac{a}{t}\right)^{1-2s}\left[F_0(s)+F_1(s)+F_2(s)\right], \tag{3.57}$$

where

$$F_0(s) \equiv \sum_{n=1}^{\infty}\frac{n^{1-s}}{(n^2+\sigma_2^2)^{s/2}}\left[F\left(s/2,1/2;s;\sigma_1^2/(n^2+\sigma_1^2)\right)-1-\frac{\sigma_1^2}{4n^2}\right],$$

$$\sigma_i^2 \equiv \frac{\lambda^{ii}\rho t^2\beta^2}{4\pi^2}, \tag{3.58}$$

$$F_1(s) \equiv \sum_{n=1}^{\infty}\frac{n^{1-s}}{(n^2+\sigma_2^2)^{s/2}}, \qquad F_2(s) \equiv \frac{\sigma_1^2}{4}\sum_{n=1}^{\infty}\frac{n^{-1-s}}{(n^2+\sigma_2^2)^{s/2}}.$$

Let us briefly study these functions. F_1 and F_2 can be analytically continued without problems:

$$F_1(s) = \zeta(2s-1) - \frac{s\sigma^2}{2}\zeta(2s+1)$$

$$+ \sum_{n=1}^{\infty}n^{1-s}\left[(n^2+\sigma_2^2)^{-s/2}-n^{-s}+\frac{s\sigma_2^2}{2}n^{-s-2}\right], \tag{3.59}$$

so that, in particular,

$$F_1(0) = -\frac{1}{12} - \frac{\sigma_2^2}{4}, \tag{3.60}$$

and

$$F_2(s) = \frac{\sigma_1^2}{4}\zeta(2s+1) + \frac{\sigma_1^2}{4}\sum_{n=1}^{\infty}n^{-1-s}\left[(n^2+\sigma_2^2)^{-s/2}-n^{-s}\right], \tag{3.61}$$

which has a pole at $s = 0$,

$$F_2(s) = \frac{\sigma_1^2}{8s} + \frac{\gamma\sigma_1^2}{4} + \mathcal{O}(s). \tag{3.62}$$

Concerning F_0, by using

$$\lim_{s \to 0} F\left(s/2, 1/2; s; \sigma_1^2/(n^2 + \sigma_1^2)\right) = \frac{1}{2}\left[1 + \sqrt{1 + \left(\frac{\sigma_1}{n}\right)^2}\right], \tag{3.63}$$

we can write

$$F_0(0) = \frac{1}{2}\sum_{n=1}^{\infty}\left[\left(n^2 + \sigma_1^2\right)^{1/2} - n - \frac{\sigma_1^2}{2n}\right]$$

$$= \frac{1}{2}\lim_{\tau \to 0}\sum_{n=1}^{\infty}\left[\left(n^2 + \sigma_1^2\right)^{1/2} - n - \frac{\sigma_1^2}{2n}\right]e^{-\tau n}. \tag{3.64}$$

This yields

$$\sum_{n=1}^{\infty} n e^{-\tau n} = \frac{1}{\tau^2} - \frac{1}{12} + \mathcal{O}(1), \tag{3.65}$$

for the term in the middle. For the other two we shall make explicit use of the results obtained in Sect. 3.1. For the last term, we get

$$\sum_{n=1}^{\infty}\frac{1}{n}e^{-\tau n} = \mathrm{Res}_{z=0}\left[\tau^{-z}\Gamma(z)\zeta(z+1)\right] + \mathcal{O}(1) = -\ln\tau + \mathcal{O}(1). \tag{3.66}$$

In order to apply this procedure to the first term we need some more specific knowledge of the function

$$\zeta_{-1/2}(z) \equiv \sum_{n=1}^{\infty} n^{-z}\left(n^2 + \sigma^2\right)^{1/2}, \tag{3.67}$$

which appears when considering (see again Sect. 3.1)

$$\sum_{n=1}^{\infty}\left(n^2 + \sigma^2\right)^{1/2} f(\tau n) = \sum_{n=1}^{\infty}\left(n^2 + \sigma^2\right)^{1/2}\frac{1}{2\pi i}\int dz\,(n\tau)^{-z}M[f](z)$$

$$= \frac{1}{2\pi i}\int dz\,\tau^{-z}M[f](z)\zeta_{-1/2}(z). \tag{3.68}$$

The study of this function can be reduced to that of Example 2. For $\mathrm{Re}\,z > -2p$ we get an analytical continuation which is a meromorphic function with poles at

$z = 2(1-k)$ of residues $\binom{1/2}{k}\sigma^{2k}$, $k = 0, 1, 2, 3, \ldots$

$$\zeta_{-1/2}^{\epsilon}(z) = \int_{\epsilon}^{\infty} dr\, r^{1-z} \left[\sqrt{1 + \frac{\sigma^2}{r^2}} - \theta(p-1) \sum_{k=0}^{p} \binom{1/2}{k} \left(\frac{\sigma}{r}\right)^{2k} \right]$$

$$+ \sum_{k=0}^{p} \binom{1/2}{k} \frac{\sigma^{2k}}{z+2(k-1)} + \epsilon^{1-z}(\epsilon^2+\sigma^2)^{1/2} - \frac{1}{2}\epsilon^{-z}(\epsilon^2+\sigma^2)^{1/2}$$

$$- \frac{i}{2\pi} \sum_{\pm} \int_{\mathbb{C}^{(\pm)}} dt\, \ln(1 - e^{\pm 2\pi i t}) \frac{d}{dt}\left[t^{-z}(t^2+\sigma^2)^{1/2} \right], \qquad (3.69)$$

where the three last terms are integer functions and the contours $\mathbb{C}^{(\pm)}$ are chosen avoiding the points $\pm i\sigma$.

In our case it is sufficient to take $p = 1$. Considering now $-2 < \operatorname{Re} z < 0$, the limit $\epsilon \to 0$ can be taken naively, with the result

$$\zeta_{-1/2}(z) = \int_{0}^{\infty} dr\, r^{1-z} \left[\sqrt{1 + \frac{\sigma^2}{r^2}} - \theta(p-1) \sum_{k=0}^{p} \binom{1/2}{k} \left(\frac{\sigma}{r}\right)^{2k} \right]$$

$$+ \sum_{k=0}^{1} \binom{1/2}{k} \frac{\sigma^{2k}}{z+2(k-1)}$$

$$- \frac{i}{2\pi} \sum_{\pm} \int_{\mathbb{C}^{(\pm)}} dt\, \ln(1 - e^{\pm 2\pi i t}) \frac{d}{dt}\left[t^{-z}(t^2+\sigma^2)^{1/2} \right]. \quad (3.70)$$

It is the last term the one which prevents continuing to the right of $z = 0$, due to its divergence at $t = 0$. This needs a special treatment (Sect. 3.1), and the final result is

$$\zeta_{-1/2}(z) = \frac{\sigma^2}{4} - \frac{\sigma^2}{2} \ln\left(\frac{\sigma}{2}\right) - \frac{\sigma}{2} + \frac{\sigma^2}{2z}$$

$$+ \frac{1}{\pi} \int_{\sigma}^{\infty} dr\, \ln(1 - e^{-2\pi r}) \frac{r}{\sqrt{r^2 - \sigma^2}} + \mathcal{O}(z). \qquad (3.71)$$

The integral term is awkward but harmless: it is exponentially suppressed as compared to the rest. Now we can go back to (3.68)

$$\sum_{n=1}^{\infty} (n^2+\sigma^2)^{1/2} e^{-\tau n} = \frac{1}{2\pi i} \int dz\, \tau^{-z} \Gamma(z) \zeta_{-1/2}(z)$$

$$= \frac{1}{\tau^2} - \frac{\sigma^2}{2} \ln\tau - \frac{\gamma\sigma^2}{2} + \frac{\sigma^2}{4} - \frac{\sigma^2}{2} \ln\left(\frac{\sigma}{2}\right) - \frac{\sigma}{2}$$

$$+ \frac{1}{\pi} \int_{\sigma}^{\infty} dr\, \ln(1 - e^{-2\pi r}) \frac{r}{\sqrt{r^2 - \sigma^2}} + \mathcal{O}_{\tau}(1). \quad (3.72)$$

And this yields for $F_0(0)$:

$$F_0(0) = -\frac{\gamma\sigma_1^2}{4} + \frac{\sigma_1^2}{8} - \frac{\sigma_1^2}{4}\ln\left(\frac{\sigma_1}{2}\right) - \frac{\sigma_1}{4} + \frac{1}{24}$$

$$+ \frac{1}{2\pi}\int_{\sigma_1}^{\infty} dr\,\ln\left(1 - e^{-2\pi r}\right)\frac{r}{\sqrt{r^2 - \sigma_1^2}}. \tag{3.73}$$

Substituting the results back into (3.57), we obtain

$$\zeta_A'(s/2) = \frac{L}{\sqrt{\pi}}\frac{\Gamma(s-1/2)}{\Gamma(s)}\left(\frac{a}{t}\right)^{1-2s}\left[\frac{\sigma_1^2}{8s} - \frac{\sigma_1^2}{8} - \frac{\sigma_1^2}{4}\ln\left(\frac{\sigma_1}{2}\right) - \frac{\sigma_1}{4} - \frac{\sigma_2^2}{4}\right.$$

$$\left. - \frac{1}{24} + \frac{1}{2\pi}\int_{\sigma_1}^{\infty} dr\,\ln\left(1 - e^{-2\pi r}\right)\frac{r}{\sqrt{r^2 - \sigma_1^2}} + \mathcal{O}(s)\right], \tag{3.74}$$

therefore

$$\zeta_A'(0) = -\frac{La\sigma_1^2}{4t} \tag{3.75}$$

and

$$\frac{d}{ds}\zeta_A'(s/2)\bigg|_{s=0} = -\frac{La}{t}\left[\psi(-1/2)\frac{\sigma_1^2}{4} - \ln(a/t)\frac{\sigma_1^2}{2} - \frac{\sigma_2^2}{2} - \frac{\sigma_1^2}{2}\ln(\sigma/2) - \frac{\sigma_1}{2}\right.$$

$$\left. + (1+\gamma)\frac{\sigma_1^2}{4} - \frac{1}{12} + \frac{1}{\pi}\int_{\sigma_1}^{\infty} dr\,\ln\left(1 - e^{-2\pi r}\right)\frac{r}{\sqrt{r^2 - \sigma_1^2}}\right]. \tag{3.76}$$

Finally

$$S_1^{(n\neq 0)} = (d-2)\frac{La}{4t}\left[\sigma_1^2 - 2\sigma_1^2\ln\left(\frac{a\sigma_1}{t}\right) - 2\sigma_2^2 - \frac{1}{3}\right.$$

$$\left. + \frac{4}{\pi}\int_{\sigma_1}^{\infty} dr\,\ln\left(1 - e^{-2\pi r}\right)\frac{r}{\sqrt{r^2 - \sigma_1^2}}\right]. \tag{3.77}$$

Adding (3.55) and (3.77) we obtain the desired expansion of the one loop effective action, S_1, near $T = 0$. The physical implications of these results have been discussed in detail in [116].

As a summary for this chapter, let us just emphasize that the series commutation techniques, that are essential in the proof of the zeta function regularization theorem—which, on its turn, is the basic tool in the general procedure of zeta function regularization when the spectrum of the operator is explicitly known (the issue in this book)—have been promoted in the above two sections to an elegant mathematical method, by making use of the Mellin transforms of convenient heat kernel operators, in combination with a rigorous treatment of the asymptotic series involved. The laborious analysis of the series to be commuted, the artificial picking up of a convenient function in order to mimic such series through pole residues on the complex plane, and the process of commutation itself, with the appearance of additional terms 'at infinity', have now disappeared, in favor of a quite natural Mellin transform analysis of the heat kernels. Moreover, the identification of the three different contributions to the final result, namely, the naive commuted series (which results in a sum of Riemann or Hurwitz zeta functions), the ordinary commutation remnants (a polynomial function) and the elusive, additional term of negative power-like behavior, appears now in a clear and natural way. The last contribution, which was originally quite difficult to handle from a numerical point of view, has been given here a completely new treatment, which allows to calculate explicit numbers with reasonable ease.

The new method has the additional advantage that it is equally well fitted for the treatment of general elliptic differential operators whose spectrum is *not* known (what is beyond the purpose of this book, see Chap. 1). The sum over eigenvalues can then be naturally replaced—within the same procedure—by a sum over heat-kernel or Seeley–De Witt coefficients. A huge mathematical industry has been generated for the calculation of these coefficients, and one can now get full profit from these result in the new context of the zeta function regularization theorem [33] (see [56, 64] for new developments).

Physical applications of these techniques keep growing, as calculations in different contexts of the vacuum energy and the Casimir effect in QFT (see Chap. 5), condensed matter and solid state physics. In the last section we have already hinted at the use one can make of those expressions for obtaining the partition functions of strings and membranes. Further results on this line will be obtained in Chap. 8.

Chapter 4
Analytical and Numerical Study of Inhomogeneous Epstein and Epstein–Hurwitz Zeta Functions

In this chapter we are going to present some of the most advanced results that we have got in our study of the different zeta functions. From the mathematical point of view they are, without any doubt, quite far reaching and involved. As will be explained in more detail later, the reason why they are not to be found in the mathematical literature dealing with zeta functions (in particular, with Epstein zeta functions) is because inhomogeneous Epstein zeta functions seem not to be very interesting in number theory—contrary to ordinary Epstein zeta functions, which are of paramount importance. The situation in physics is just the opposite: ordinary Epstein zeta functions appear as a very limited particular case—massless, zero temperature, no chemical potential—of the usual theories.

The formula (4.32) that will be obtained in Sect. 4.1 is due to the author. It constitutes an original and non-trivial extension of the celebrated Chowla–Selberg formula [5, 25]. Surprisingly enough, the new formula—which solves the non-homogeneous case, the most important in physics—turns out to be in the end as beautiful (mathematically) as the one derived by those famous mathematicians for the homogeneous situation. To see the importance that Chowla and Selberg attributed to their discovery, the reader is advised to throw a look at Ref. [25]. It goes without saying that the author is equally happy with the new formula (4.32).

For physical applications, the importance of the formula—and of the other ones obtained in the present and in the preceding chapters—resides in the quick convergence (of exponential type) of the series appearing in them (see again [25], for the usefulness of this property). This fact will be illustrated in detail in Sect. 4.2 of the present chapter, with specific numerical calculations carried out for the most useful of such formulas. Explicit physical applications will start from the next chapter.

E. Elizalde, *Ten Physical Applications of Spectral Zeta Functions*,
Lecture Notes in Physics 855,
DOI 10.1007/978-3-642-29405-1_4, © Springer-Verlag Berlin Heidelberg 2012

4.1 Explicit Analytical Continuation of Inhomogeneous Epstein Zeta Functions

In this section the two-dimensional inhomogeneous zeta-function series (with homogeneous part of the most general Epstein type):

$$\sum_{m,n\in\mathbb{Z}}{}'\left(am^2 + bmn + cn^2 + q\right)^{-s},\tag{4.1}$$

will be analytically continued in the variable s. The result shows the pole structure explicitly, and is given in terms of a convergent series in one index only. This extends a previous formula by S. Chowla and A. Selberg, while preserving for any $q > 0$ the good convergence properties of the final series, which led these authors to obtain very important applications of the formula in number theory and in the theory of elliptic functions. Further applications to several unsolved problems of mathematical physics as well as some direct physical applications of the formula will be described too.

In a paper by the mathematicians S. Chowla and A. Selberg published some years ago [25], these authors considered several applications of an interesting formula they had found for the analytical continuation of the general Epstein zeta function in two indices [2, 4]

$$\sum_{m,n\in\mathbb{Z}}{}'\left(am^2 + bmn + cn^2\right)^{-s}\tag{4.2}$$

(the prime means, as usually, that the term with $m = n = 0$ is to be excluded from the sum). This formula was written in that article without any explicit derivation or hint of any kind (the actual derivation was promised to appear in a subsequent paper), and became to be named after their authors (see, for instance, Ref. [5]). As remarked by Chowla and Selberg themselves, the good convergence properties of the series of Bessel functions which appears in the formula, make it both simple to apply and very useful. In particular, it was employed in [25] to construct an easy proof of the famous conjecture by Gauss on the class-number of binary quadratic forms with a negative fundamental discriminant (an alternative to the derivation of Heilbronn [117], which was based on earlier work by Deuring [118]), to demonstrate the positiveness of certain Dirichlet L-functions at $s = 1/2$, and also in a classical problem of the theory of elliptic functions, where the range of computability in finite terms of the standard K elliptic integral was extended considerably.

Guided by these results and by the necessity to obtain a similar formula for the case of the 'Epstein like' zeta function corresponding to a quadratic form plus a constant term, we will here consider the double, doubly infinite series

$$E(s; a, b, c; q) \equiv \sum_{m,n\in\mathbb{Z}}{}'\left(am^2 + bmn + cn^2 + q\right)^{-s}.\tag{4.3}$$

With $q \neq 0$, in general. The parenthesis in (4.3) is the inhomogeneous quadratic form

$$Q(x, y) + q, \qquad Q(x, y) \equiv ax^2 + bxy + cy^2, \tag{4.4}$$

restricted to the integers. In the general theory that deals with the homogeneous case, one assumes that $a, c > 0$ and that the discriminant

$$\Delta = 4ac - b^2 > 0 \tag{4.5}$$

(see Ref. [25]). Here we will impose the additional condition that q be such that $Q(m, n) + q \neq 0$, for all $m, n \in \mathbb{Z}$. In the usual applications of the theory, those conditions are indeed satisfied.

Before we attack this problem, and as a more simple application of the general formalism, we will first consider the case (also interesting for its many applications) of the analytical continuation of the more simple series

$$G(s; a, c; q) \equiv \sum_{n=-\infty}^{+\infty} \left[a(n + c)^2 + q \right]^{-s}. \tag{4.6}$$

The result for this case can be considered as a byproduct of our general approach. This calculation will be carried out in Sect. 4.1 of the chapter. For the benefit of the reader, the derivation of the extended formula for the general case will be given in detail.

4.1.1 The Particular Case of the Basic One-Dimensional Epstein–Hurwitz Series

While deriving this case we shall already introduce the general procedure. Actually, the analytical continuation can be performed in different ways, but maybe the most direct method is by using Jacobi's identity for the theta function θ_3 (see also Chap. 2)

$$\theta_3(z|\tau) = 1 + 2 \sum_{n=1}^{\infty} q^{n^2} \cos(2nz), \quad q = e^{\pi i \tau}, \ |q| < 1, \ \tau \in \mathbb{C}, \tag{4.7}$$

that is

$$\theta_3(z|\tau) = \frac{1}{\sqrt{-i\tau}} e^{z^2/(\pi i \tau)} \theta_3\left(\frac{z}{\tau} \,\middle|\, \frac{-1}{\tau} \right), \tag{4.8}$$

or, in other words

$$\sum_{n=-\infty}^{+\infty} e^{n^2 \pi i \tau + 2niz} = \frac{1}{\sqrt{-i\tau}} \sum_{n=-\infty}^{+\infty} e^{(z-n\pi)^2/(\pi i \tau)} \tag{4.9}$$

(for a popular reference see, for instance, Wittaker and Watson [119, p. 476]). Here z and τ are arbitrary complex, $z, \tau \in \mathbb{C}$, with the only restriction that $\operatorname{Im} \tau > 0$ (in order that $|q| < 1$). For subsequent application, it turns out to be better to recast the Jacobi identity as follows (with $\pi i \tau \to -t$ and $z \to \pi z$):

$$\sum_{n=-\infty}^{+\infty} e^{-n^2 t + 2\pi i n z} = \sqrt{\frac{\pi}{t}} \sum_{n=-\infty}^{+\infty} e^{-\pi^2 (n-z)^2 / t}, \tag{4.10}$$

equivalently

$$\sum_{n=-\infty}^{+\infty} e^{-(n+z)^2 t} = \sqrt{\frac{\pi}{t}} \left[1 + \sum_{n=1}^{\infty} e^{-\pi^2 n^2 / t} \cos(2\pi n z) \right], \tag{4.11}$$

where $z, t \in \mathbb{C}$, $\operatorname{Re} t > 0$. This last expression will be the first ingredient in our calculation. Instead, we could have chosen to proceed by expanding in power series the exponent of the Mellin transform of the series above, and then by interchanging the order of the summations of the two series, with the known prescription of adding the contribution of the corresponding contour integration on the complex plane— see Chap. 2 and Refs. [48, 51, 109], where this general method has been extensively applied (even in the case of arbitrary, non-negative exponents), and was shown to yield the Jacobi identity as a particular case (see also Refs. [10, 11, 14, 120] for more extensive accounts). Actually, that both procedures should yield the same result in the quadratic case is well known from the direct proof given by Landsberg [121] of the Jacobi identity.

Another ingredient for our derivation is the gamma function identity. Applied to the abbreviated expression $\sum (Q + q)^{-s}$ (which precise meaning corresponding to (4.3) or (4.6) can be left here undefined, for the moment), it yields

$$\sum (Q + q)^{-s} = \frac{1}{\Gamma(s)} \sum \int_0^\infty du \, u^{s-1} e^{-(Q+q)u}. \tag{4.12}$$

When $q > 0$ and under the conditions that have been imposed to the quadratic form Q, the integral and the sum commute. Before considering the general case, by setting first $Q = a(n + c)^2$, $a > 0$, $c \in \mathbb{C}$, $c \neq 0, -1, -2, \ldots$ (of course, this is not a quadratic form and the series is now one-dimensional, but the same considerations about commutation of series and integral apply), we obtain the following expression (a consequence of the Jacobi and gamma function identity)

$$\sum_{n=-\infty}^{+\infty} \left[a(n+c)^2 + q \right]^{-s} = \sqrt{\frac{\pi}{a}} \frac{\Gamma(s - 1/2)}{\Gamma(s)} q^{1/2-s} + \frac{4\pi^s}{\Gamma(s)} a^{-1/4-s/2} q^{1/4-s/2}$$

$$\cdot \sum_{n=1}^{\infty} n^{s-1/2} \cos(2\pi n c) K_{s-1/2}(2\pi n \sqrt{q/a}), \tag{4.13}$$

where K_ν is the modified Bessel function of the second kind. This equation provides the analytic continuation of what can be named the most basic inhomogeneous, generalized Epstein series in one index, in terms of a series that turns out to be quickly convergent for extensive ranges of values of the parameters, in particular, when $\mathrm{Im}\, c < \sqrt{q/a}$, a case that frequently appears in the applications. An example of a result that can only be obtained by explicit, non-trivial interchange of the summation indices and subsequent contour integration—and that it is *not* a consequence of the Jacobi identity—is the following:

$$
F(s; a, c; q)
$$

$$
\equiv \sum_{n=0}^{\infty} [a(n+c)^2 + q]^{-s}
$$

$$
\sim \frac{q^{-s}}{\Gamma(s)} \sum_{m=0}^{\infty} \frac{(-1)^m \Gamma(m+s)}{m!} \left(\frac{q}{a}\right)^{-m} \zeta_H(-2m, c)
$$

$$
+ \sqrt{\frac{\pi}{a}} \frac{\Gamma(s-1/2)}{2\Gamma(s)} q^{1/2-s}
$$

$$
+ \frac{2\pi^s}{\Gamma(s)} a^{-1/4-s/2} q^{1/4-s/2} \sum_{n=1}^{\infty} n^{s-1/2} \cos(2\pi nc) K_{s-1/2}(2\pi n\sqrt{q/a}).
$$

$$(4.14)$$

This function will be studied in detail in the next section (see also Ref. [122]) with numerical tables, plots, and a couple of explicit applications. Notice, however, that this is not a convergent series but an asymptotic one [48, 51, 109]. By observing that

$$
\zeta_H(-2m, c) + \zeta_H(-2m, 1-c) = 0, \tag{4.15}
$$

for $m \in \mathbb{N}$, and that

$$
G(s; a, c; q) = F(s; a, c; q) + F(s; a, 1-c; q), \tag{4.16}
$$

it is quite easy to obtain (4.13) as a particular case of (4.14).

Those formulas give explicit answers to some questions that had remained unsolved in zeta-function regularization. Only the case $c = 0$ had been dealt with satisfactorily (through the corresponding expression obtained just putting $c = 0$ in (2.93)). As a particular case of (2.93), we obtain

$$
\sum_{n=0}^{\infty} [a(n+1/2)^2 + q]^{-s} = \frac{1}{2} \sqrt{\frac{\pi}{a}} \frac{\Gamma(s-1/2)}{\Gamma(s)} q^{1/2-s} + \frac{2\pi^s}{\Gamma(s)} a^{-1/4-s/2} q^{1/4-s/2}
$$

$$
\cdot \sum_{n=1}^{\infty} (-1)^n n^{s-1/2} K_{s-1/2}(2\pi n\sqrt{q/a}). \tag{4.17}
$$

More involved multiple series of this kind, such as the diagonal, inhomogeneous, generalized Epstein (or Epstein–Hurwitz) multiple series

$$E_k(s; a_1, \ldots, a_k; c_1, \ldots, c_k; c)$$

$$\equiv \sum_{n_1, \ldots, n_k \in \mathbb{Z}} \left[a_1(n_1 + c_1)^2 + \cdots + a_k(n_k + c_k)^2 + c \right]^{-s}, \qquad (4.18)$$

can be treated in a recurrent way, starting from (4.13). The general recurrence is (see also Chap. 2)

$$E_k(s; a_1, \ldots, a_k; c_1, \ldots, c_k; c)$$

$$= \sqrt{\frac{\pi}{a_k}} \frac{\Gamma(s - 1/2)}{\Gamma(s)} E_{k-1}(s; a_1, \ldots, a_{k-1}; c_1, \ldots, c_{k-1}; c)$$

$$+ \frac{4\pi^s}{\Gamma(s)} a_k^{-s/2-1/4} \sum_{n_1, \ldots, n_{k-1} \in \mathbb{Z}} \left[\sum_{j=1}^{k-1} a_j(n_j + c_j)^2 + c \right]^{-s/2+1/4}$$

$$\cdot \sum_{n_k=1}^{\infty} n_k^{s-1/2} \cos(2\pi n_k c_k) K_{s-1/2}\left(\frac{2\pi n_k}{\sqrt{a_k}} \sqrt{\sum_{j=1}^{k-1} a_j(n_j + c_j)^2 + c} \right).$$

$$(4.19)$$

This recurrence is very appropriate for numerical computation, since from the second term on the r.h.s., owing to the rapid exponential convergence of the Bessel function, only a few first terms need to be taken into account to achieve a good approximation. The recurrence is then implementable in any of the algebraic computational packages commonly available.

4.1.2 The Homogeneous Case: Chowla–Selberg's Formula

Here, the Chowla–Selberg formula [25] for the (general homogeneous) Epstein zeta function [2, 4] corresponding to the quadratic form Q is to be used. (This is an expression well known in number theory [5] but not so much in mathematical physics.) The result is

$$F(s; a, b, c; 0) = 2\zeta(2s)a^{-s} + \frac{2^{2s}\sqrt{\pi}a^{s-1}}{\Gamma(s)\Delta^{s-1/2}} \Gamma(s - 1/2)\zeta(2s - 1)$$

$$+ \frac{2^{s+3/2}\pi^s}{\Gamma(s)\Delta^{s/2-1/4}\sqrt{a}} \sum_{n=1}^{\infty} n^{s-1/2}\sigma_{1-2s}(n)\cos(n\pi b/a)$$

$$\cdot \int_0^{\infty} dt\, t^{s-3/2} \exp\left[-\frac{\pi n\sqrt{\Delta}}{2a}(t + t^{-1}) \right], \qquad (4.20)$$

where

$$\sigma_s(n) \equiv \sum_{d|n} d^s, \tag{4.21}$$

namely the sum over the s-powers of the divisors of n. Notice that there is an error in the transcription of the formula in [5].

This formula is very useful and its practical application quite simple. In fact, the two first terms are just nice, while the last one is quickly convergent and thus causes no problem: only a few first terms of the series need to be calculated, even if one needs exceptionally good accuracy. One should also notice that the pole of F at $s = 1$ appears through $\zeta(2s-1)$ in the second term, while for $s = 1/2$, the apparent singularities of the first and second terms cancel each other and no pole is formed.

A closer, quantitative idea about the integral can be got from the following closed expression for it:

$$\int_0^\infty dt\, t^{\nu-1} \exp\left(-\frac{\alpha}{t} - \beta t\right) = 2\left(\frac{\alpha}{\beta}\right)^{\nu/2} K_\nu(2\sqrt{\alpha\beta}), \tag{4.22}$$

K_ν being again the modified Bessel function of the second kind. In particular, by calling the integral

$$I(n, s) \equiv \int_0^\infty dt\, t^{s-3/2} \exp\left[-\frac{\pi n\sqrt{\Delta}}{2a}(t + t^{-1})\right], \tag{4.23}$$

one has

$$I(n, 0) = \sqrt{\frac{2a}{n\sqrt{\Delta}}}\, \exp\left(-\frac{\pi n\sqrt{\Delta}}{a}\right) = I(n, 1),$$

$$I(n, 1/2) = 2K_0(\pi n\sqrt{\Delta}/a),$$

$$I(n, 2) = \frac{a + \pi n\sqrt{\Delta}}{\pi n\sqrt{\Delta}} \sqrt{\frac{2a}{n\sqrt{\Delta}}}\, \exp\left(-\frac{\pi n\sqrt{\Delta}}{a}\right), \tag{4.24}$$

$$I(n, 3) = \frac{3a^2 + 3\pi na\sqrt{\Delta} + \pi^2 n^2 \Delta}{\pi^2 n^2 \Delta} \sqrt{\frac{2a}{n\sqrt{\Delta}}}\, \exp\left(-\frac{\pi n\sqrt{\Delta}}{a}\right).$$

As functions of n, all these expressions share the common feature of being exponentially decreasing with n.

4.1.3 Derivation of the General Two-Dimensional Formula

To deal with the case when Q is a two-dimensional quadratic form and the sum $\sum$ a double, doubly infinite series, we need a further ingredient, i.e., casting the quadratic form $Q(n_1, n_2)$ as the sum of two squares

$$Q(n_1, n_2) = a\left[\left(n_1 + \frac{bn_2}{2a}\right)^2 + \frac{\Delta}{4a^2}n_2^2\right], \qquad (4.25)$$

and then to proceed by considering first the summation over n_1 (this will be the sum to which the Jacobi identity (4.10) will be applied), while treating n_2 as a parameter. Doing so, performing then the change of variables

$$u = \frac{2\pi n_1}{\sqrt{\Delta}\, n_2}t, \qquad (4.26)$$

and taking advantage of the common idea of rewriting the double sum as a sum over the product $n = n_1 n_2$ and (a finite one) over the divisors of the product:

$$\sum_{n_1,n_2}\left(\frac{n_1}{n_2}\right)^{s-1/2} = \sum_{n_1,n_2}(n_1 n_2)^{s-1/2}n_2^{1-2s} = \sum_n n^{s-1/2}\sum_{d|n}d^{1-2s}, \qquad (4.27)$$

the following expression is obtained

$$E(s; a, b, c; q)$$

$$= {\sum_{m,n\in\mathbb{Z}}}'\left[Q(m,n)+q\right]^{-s}$$

$$= {\sum_{m,n\in\mathbb{Z}}}'\left(am^2 + bmn + cn^2 + q\right)^{-s}$$

$$= 2\zeta_{EH}(s, q/a)a^{-s} + \frac{2^{2s}\sqrt{\pi}\,a^{s-1}}{\Gamma(s)\Delta^{s-1/2}}\Gamma(s-1/2)\zeta_{EH}(s-1/2, 4aq/\Delta)$$

$$+ \frac{2^{s+3/2}\pi^s}{\Gamma(s)\Delta^{s/2-1/4}\sqrt{a}}\sum_{n=1}^{\infty}n^{s-1/2}\cos(n\pi b/a)\sum_{d|n}d^{1-2s}\int_0^\infty dt\, t^{s-3/2}$$

$$\cdot \exp\left\{-\frac{\pi n\sqrt{\Delta}}{2a}\left[\left(1 + \frac{4aq}{d^2\Delta}\right)t + t^{-1}\right]\right\}, \qquad (4.28)$$

where the function $\zeta_{EH}(s, p)$ (the one-dimensional Epstein–Hurwitz or inhomogeneous Epstein series) is given by

$$\zeta_{EH}(s; p) = \sum_{n=1}^{\infty}(n^2 + p)^{-s} = \frac{1}{2}{\sum_{n\in\mathbb{Z}}}'(n^2 + p)^{-s}$$

$$= -\frac{p^{-s}}{2} + \frac{\sqrt{\pi}\Gamma(s-1/2)}{2\Gamma(s)}p^{-s+1/2}$$

$$+ \frac{2\pi^s p^{-s/2+1/4}}{\Gamma(s)}\sum_{n=1}^{\infty}n^{s-1/2}K_{s-1/2}(2\pi n\sqrt{p}), \qquad (4.29)$$

as a particular case of (4.13) (or (4.14)).

It is remarkable that the integral inside the series of the new expression can still be written in a closed form using (4.22)—as in the case of (4.20). Calling now the integral

$$J(n,s) \equiv \int_0^\infty dt\, t^{s-3/2} \exp\left\{ -\frac{\pi n \sqrt{\Delta}}{2a}\left[\left(1 + \frac{4aq}{\Delta d^2}\right)t + t^{-1} \right] \right\}, \qquad (4.30)$$

we obtain, in particular,

$$J(n,0) = \sqrt{\frac{2a}{n\sqrt{\Delta}}}\, \exp\left[-\frac{\pi n}{a}\left(\Delta + \frac{4aq}{d^2}\right)^{1/2} \right],$$

$$J(n,1/2) = 2K_0\left(\frac{\pi n}{a}\sqrt{\Delta + \frac{4aq}{d^2}} \right),$$

$$J(n,1) = \sqrt{\frac{2a\sqrt{\Delta}}{n}}\left(\Delta + \frac{4aq}{d^2}\right)^{-1/2} \exp\left[-\frac{\pi n}{a}\left(\Delta + \frac{4aq}{d^2}\right)^{1/2} \right],$$

$$J(n,2) = \left(\frac{a}{\pi n} + \sqrt{\Delta + \frac{4aq}{d^2}} \right)\sqrt{\frac{2a\Delta^{3/2}}{n}}\left(\Delta + \frac{4aq}{d^2}\right)^{-3/2} \qquad (4.31)$$

$$\cdot \exp\left[-\frac{\pi n}{a}\left(\Delta + \frac{4aq}{d^2}\right)^{1/2} \right],$$

$$J(n,3) = \left(\frac{3a^2}{\pi^2 n^2} + \frac{3a}{\pi n}\sqrt{\Delta + \frac{4aq}{d^2}} + \Delta + \frac{4aq}{d^2} \right)\sqrt{\frac{2a\Delta^{5/2}}{n}}$$

$$\cdot \left(\Delta + \frac{4aq}{d^2}\right)^{-5/2} \exp\left[-\frac{\pi n}{a}\left(\Delta + \frac{4aq}{d^2}\right)^{1/2} \right],$$

which are again exponentially decreasing with n.

Expression (4.28)—as (4.13)—can be then written in terms of modified Bessel functions of the second kind, to yield finally

$$E(s; a, b, c; q)$$

$$= 2\zeta_{EH}(s, q/a)a^{-s} + \frac{2^{2s}\sqrt{\pi}a^{s-1}}{\Gamma(s)\Delta^{s-1/2}}\Gamma(s-1/2)\zeta_{EH}(s-1/2, 4aq/\Delta)$$

$$+ \frac{2^{s+5/2}\pi^s}{\Gamma(s)\sqrt{a}} \sum_{n=1}^\infty n^{s-1/2}\cos(n\pi b/a)$$

$$\cdot \sum_{d|n} d^{1-2s}\left(\Delta + \frac{4aq}{d^2}\right)^{1/4-s/2} K_{s-1/2}\left(\frac{\pi n}{a}\sqrt{\Delta + \frac{4aq}{d^2}} \right). \qquad (4.32)$$

(4.32) provides the analytical continuation of the inhomogeneous Epstein series, in the variable s, as a meromorphic function in the complex plane. Its pole structure is explicitly given in terms of the well-known pole structure of $\zeta_{EH}(s, p)$ (see, for instance, [122]).

(4.32) constitutes a major result of this chapter. We propose to call it the *extended Chowla–Selberg* formula, since it certainly contains the Chowla–Selberg formula as the particular case $q = 0$, i.e.,

$$E(s; a, b, c; 0)$$

$$= 2\zeta(2s)a^{-s} + \frac{2^{2s}\sqrt{\pi}\,a^{s-1}}{\Gamma(s)\Delta^{s-1/2}}\Gamma(s - 1/2)\zeta(2s - 1) + \frac{2^{s+5/2}\pi^s}{\Gamma(s)\Delta^{s/2-1/4}\sqrt{a}}$$

$$\cdot \sum_{n=1}^{\infty} n^{s-1/2}\sigma_{1-2s}(n)\cos(n\pi b/a)K_{s-1/2}\left(\frac{\pi n\sqrt{\Delta}}{a}\right). \tag{4.33}$$

Formula (4.32) had never appeared in the mathematical literature. The good convergence properties of expression (4.33), that were so much prised by Chowla and Selberg, are shared by its non-trivial extension (4.32). This renders the use of the formula quite simple. In fact, the two first terms are still rather nice—under the form (4.29)—while the last one (impressive in appearance) is even more quickly convergent than in the case of (4.33), and thus absolutely harmless in fact. Only a few first terms of the three series of Bessel functions in (4.32), (4.29) need to be calculated, even for obtaining very good accuracy. Also here, the pole of $E(s)$, (4.33), at $s = 1$ appears through the $\zeta(2s - 1)$ in the second term, while for $s = 1/2$, the apparent singularities of the first and second terms cancel each other yielding a finite contribution. Analogously, the pole at $s = 1/2$ in (4.32) appears only from the first term. It is remarkable that (4.32) possesses also these good properties for any non-negative value of q. In fact, for large q the convergence properties of the series of Bessel functions are clearly enhanced, while for q small we get back to the case of Chowla and Selberg. Notice, however, that this limit is *not* obtained through the high-q expansion, (4.14), but, on the contrary, using the low-q, binomial expansion:

$$\sum_{n=0}^{\infty}\left[a(n + c)^2 + q\right]^{-s}$$

$$= a^{-s}\sum_{m=0}^{\infty}\frac{(-1)^m\Gamma(m + s)}{\Gamma(s)m!}\left(\frac{q}{a}\right)^m\zeta_H(2s + 2m, c), \tag{4.34}$$

which is convergent for $q/a \leq 1$. For $q \to 0$ it reduces to $a^{-s}\zeta_H(2s, c)$. Actually, formula (4.32) is still valid in a domain of negative q's, namely for $q > -\min(a, c, a - b + c)$.

4.2 Extended Chowla–Selberg Formulas, Associated with Arbitrary Forms of Quadratic+Linear+Constant Type

Consider now the following zeta function ($\operatorname{Re} s > p/2$):

$$\zeta_{A,\vec{c},q}(s) = {\sum_{\vec{n}\in\mathbb{Z}^p}}' \left[\frac{1}{2}(\vec{n}+\vec{c})^T A(\vec{n}+\vec{c}) + q \right]^{-s} \equiv {\sum_{\vec{n}\in\mathbb{Z}^p}}' [Q(\vec{n}+\vec{c}) + q]^{-s}. \quad (4.35)$$

The prime on a summation sign means that the point $\vec{n} = \vec{0}$ is to be excluded from the sum. As we shall see, this is irrelevant when q or some component of $\vec{c}$ is non-zero but, on the contrary, it becomes an inescapable condition in the case when $c_1 = \cdots = c_p = q = 0$. Note that, alternatively, we can view the expression inside the square brackets of the zeta function as a sum of a quadratic, a linear, and a constant form, namely, $Q(\vec{n}+\vec{c}) + q = Q(\vec{n}) + L(\vec{n}) + \bar{q}$.

Our aim is to obtain a formula that gives (the analytic continuation of) this multidimensional zeta function in terms of an exponentially convergent series, and which is valid in the whole complex plane, exhibiting the singularities (poles) of the meromorphic continuation—with the corresponding residua—explicitly. The only condition on the matrix A is that it correspond to a (non-negative) quadratic form, which we call Q. The vector $\vec{c}$ is arbitrary, while q will be (for the moment) a positive constant. As we shall see, the solution to this problem will depend very much (its explicit form) on the fact that q and/or $\vec{c}$ are zero or not. According to this, we will have to distinguish different cases, leading to unrelated final formulas, all to be viewed as different non-trivial extensions of the CS formula (they will be named ECS formulas, and will carry additional tags, for the different cases).

Use of the Poisson resummation formula in (4.35) yields [123–125]

$$\zeta_{A,\vec{c},q}(s) = \frac{(2\pi)^{p/2}q^{p/2-s}}{\sqrt{\det A}} \frac{\Gamma(s-p/2)}{\Gamma(s)} + \frac{2^{s/2+p/4+2}\pi^s q^{-s/2+p/4}}{\sqrt{\det A}\,\Gamma(s)}$$

$$\cdot {\sum_{\vec{m}\in\mathbb{Z}^p_{1/2}}}' \cos(2\pi\vec{m}\cdot\vec{c})\,(\vec{m}^T A^{-1}\vec{m})^{s/2-p/4} K_{p/2-s}\left(2\pi\sqrt{2q\vec{m}^T A^{-1}\vec{m}}\right),$$

$$(4.36)$$

where K_ν is the modified Bessel function of the second kind and the subindex $1/2$ in $\mathbb{Z}^p_{1/2}$ means that in this sum, only half of the vectors $\vec{m} \in \mathbb{Z}^p$ enter. That is, if we take an $\vec{m} \in \mathbb{Z}^p$ we must then exclude $-\vec{m}$ (as a simple criterion one can, for instance, select those vectors in $\mathbb{Z}^p\setminus\{\vec{0}\}$ whose first non-zero component is positive). (4.36) fulfills *all* the requirements of a CS formula. But it is very different from the original one, constituting a non-trivial extension to the case of a quadratic+linear+constant form, in any number of dimensions, with the constant term being non-zero. We shall denote this formula, (4.36), by the acronym ECS1.

It is notorious how the only pole of this inhomogeneous Epstein zeta function appears, explicitly, at $s = p/2$, where it belongs. Its residue is given by:

$$\text{Res}_{s=p/2}\,\zeta_{A,\vec{c},q}(s) = \frac{(2\pi)^{p/2}}{\Gamma(p/2)}(\det A)^{-1/2}. \tag{4.37}$$

4.2.1 Limit $q \to 0$

After some work, one can obtain the limit of expression (4.36) as $q \to 0$ (for simplicity we also set $\vec{c} = \vec{0}$)

$$\zeta_{A,\vec{0},0}(s) = 2^{1+s}a^{-s}\zeta(2s) + \sqrt{\frac{\pi}{a}}\frac{\Gamma(s-1/2)}{\Gamma(s)}\zeta_{A_{p-1},\vec{0},0}(s-1/2) + \frac{4\pi^s}{a^{s/2+1/4}\Gamma(s)}$$

$$\cdot \sum_{\vec{n}_2\in\mathbb{Z}^{p-1}}{}'\sum_{n_1=1}^{\infty}\cos\left(\frac{\pi n_1}{a}\vec{b}^T\vec{n}_2\right)n_1^{s-1/2}\left(\vec{n}_2^T\Delta_{p-1}\vec{n}_2\right)^{1/4-s/2}$$

$$\cdot K_{s-1/2}\left(\frac{2\pi n_1}{\sqrt{a}}\sqrt{\vec{n}_2^T\Delta_{p-1}\vec{n}_2}\right). \tag{4.38}$$

In (4.36) and (4.38), A is a $p \times p$ symmetric matrix $A = (a_{ij})_{i,j=1,2,\dots,p} = A^T$, A_{p-1} the $(p-1) \times (p-1)$ reduced submatrix $A_{p-1} = (a_{ij})_{i,j=2,\dots,p}$, a the first component, $a = a_{11}$, $\vec{b}$ the $p-1$ vector $\vec{b} = (a_{21},\dots,a_{p1})^T = (a_{12},\dots,a_{1p})^T$, and Δ_{p-1} is the following $(p-1) \times (p-1)$ matrix: $\Delta_{p-1} = A_{p-1} - \frac{1}{4a}\vec{b}\otimes\vec{b}$. More precisely, what one actually obtains by taking the limit is *the reflected formula*, as one would get after using the Epstein zeta function reflection $\Gamma(s)Z(s;A) = \pi^{2s-p/2}(\det A)^{-1/2}\Gamma(p/2-s)Z(p/2-s;A^{-1})$, being $Z(s;A)$ the Epstein zeta function [2, 4]. Finally. it can be written as (4.38). (It is a rather non-trivial exercise to perform this calculation.) Note that (4.38) has *all* the properties demanded from a CS formula, but it is actually *not explicit*. It is in fact a recurrence, rather lengthy to solve as it stands. In fact, it can be viewed as the *straightforward extension* of the original CS formula to higher dimensions. It was the top result of previous work on this subject, for the case $q = c_1 = \dots = c_p = 0$ [123–125].

Using a different strategy, this recurrence will be now solved explicitly, in a much more simple way. Indeed, let us proceed in a complementary way, namely, by doing the inversion provided by the Poisson resummation formula (or the Jacobi identity), with respect to $p-1$ of the indices (say, $j = 2, 3, \dots, p$). This leaves us with three sums, corresponding to positive, zero, and negative values of the remaining index (n_1, in this case). The zero value of n_1 (in correspondence with the rest of the n_i's not being all zero) classifies the number of different situations (according to the values of the c_i's an q being all zero or not) into just two cases. (As is immediate, from start all c_i's can be taken to be between 0 and 1: $0 \le c_i < 1$, $i = 1, 2, \dots, p$.) (i) The first case is, thus, when at least one of the c_i's or $q \ge 0$ is not zero. Since the

case $q \neq 0$ has been solved already, we will mean by this case now that, say $c_1 \neq 0$.
(ii) The second case is when all $q = c_1 = \cdots = c_p = 0$.

4.2.2 Case with $q = 0$ but $c_1 \neq 0$

General (Non-diagonal) Subcase   By doing the inversion provided by the Poisson resummation formula (or the Jacobi identity), with respect to $p - 1$ of the indices (here, $j = 2, 3, \ldots, p$), we readily obtain:

$$
\zeta_{A_p, \vec{c}, 0}(s)
$$

$$
= \frac{2^s}{\Gamma(s)} (\det A_{p-1})^{-1/2} \Bigg\{ \pi^{(p-1)/2} \big(a_{11} - \vec{a}_{p-1}^T A_{p-1}^{-1} \vec{a}_{p-1} \big)^{(p-1)/2-s}
$$

$$
\cdot \Gamma\big(s - (p-1)/2\big) \big[\zeta_H(2s - p + 1, c_1) + \zeta_H(2s - p + 1, 1 - c_1) \big]
$$

$$
+ 4\pi^s \big(a_{11} - \vec{a}_{p-1}^T A_{p-1}^{-1} \vec{a}_{p-1} \big)^{(p-1)/4-s/2}
$$

$$
\cdot \sum_{n_1 \in \mathbb{Z}} \sum_{\vec{m} \in \mathbb{Z}_{1/2}^{p-1}} {}' \cos\big[2\pi \vec{m}^T \big(\vec{c}_{p-1} + A_{p-1}^{-1} \vec{a}_{p-1}(n_1 + c_1) \big) \big]
$$

$$
\cdot |n_1 + c_1|^{(p-1)/2-s} \big(\vec{m}^T A_{p-1}^{-1} \vec{m} \big)^{s/2-(p-1)/4}
$$

$$
\cdot K_{(p-1)/2-s}\Big(2\pi |n_1 + c_1| \sqrt{\big(a_{11} - \vec{a}_{p-1}^T A_{p-1}^{-1} \vec{a}_{p-1} \big) \vec{m}^T A_{p-1}^{-1} \vec{m}} \ \Big) \Bigg\}
$$

$$
- \Big(\tfrac{1}{2} \vec{c}^T A \vec{c} \Big)^{-s}. \tag{4.39}
$$

Here, and in what follows, A_{p-1} is (as before) the submatrix of A_p composed of the last $p - 1$ rows and columns. Moreover, a_{11} is the first diagonal component of A_p, while $\vec{a}_{p-1} = (a_{12}, \ldots, a_{1p})^T = (a_{21}, \ldots, a_{p1})^T$, and $\vec{m} = (n_2, \ldots, n_p)^T$. Note that this is an *explicit formula*, that the only pole at $s = p/2$ appears also explicitly, and that the second term of the rhs is a series of exponentially fast convergence. It has, therefore (as (4.36)), all the properties required to qualify as a CS formula. We shall name this expression ECS2.

Diagonal Subcase   In this very common, particular case the preceding expression reduces to the more simple form:

$$
\zeta_{A_p, \vec{c}, 0}(s) = \frac{2^s}{\Gamma(s)} (\det A_{p-1})^{-1/2} \Big\{ \pi^{(p-1)/2} a_1^{(p-1)/2-s} \Gamma\big(s - (p-1)/2\big)
$$

$$
\cdot \big[\zeta_H(2s - p + 1, c_1) + \zeta_H(2s - p + 1, 1 - c_1) \big]
$$

$$+ 4\pi^s a_1^{(p-1)/4 - s/2} \sum_{n_1 \in \mathbb{Z}} \sum_{\vec{m} \in \mathbb{Z}_{1/2}^{p-1}}{}' \cos\left(2\pi \vec{m}^T \vec{c}_{p-1}\right)$$

$$\cdot |n_1 + c_1|^{(p-1)/2 - s} \left(\vec{m}^T A_{p-1}^{-1} \vec{m}\right)^{s/2 - (p-1)/4}$$

$$\cdot K_{(p-1)/2 - s}\left(2\pi |n_1 + c_1| \sqrt{a_1 \vec{m}^T A_{p-1}^{-1} \vec{m}}\right)\Bigg\} - \left(\frac{1}{2}\vec{c}^T A \vec{c}\right)^{-s}. \quad (4.40)$$

We shall call this formula ECS2d.

4.2.3 Case with $c_1 = \cdots = c_p = q = 0$

General (Non-diagonal) Subcase   As remarked in [123–125], we had not been able to obtain here yet a closed formula, but just a (rather non-trivial) recurrence, (4.38), relating the p-dimensional case with the $(p-1)$-dimensional one. After a second look, we have now realized that we can actually still proceed as if we had in fact $c_1 = 1 \neq 0$, both for positive and for negative values of n_1. A sum, though, remains with $n_1 = 0$—and the rest of the n_i's not all being zero—what yields, once more, the same zeta function of the beginning, but corresponding to $p-1$ indices. All in all:

$$\zeta_{A_p, \vec{0}, 0}(s) = \zeta_{A_{p-1}, \vec{0}, 0}(s) + \frac{2^{1+s}}{\Gamma(s)} (\det A_{p-1})^{-1/2}$$

$$\cdot \left\{ \pi^{(p-1)/2} \left(a_{11} - \vec{a}_{p-1}^T A_{p-1}^{-1} \vec{a}_{p-1}\right)^{(p-1)/2 - s} \right.$$

$$\cdot \Gamma\left(s - (p-1)/2\right) \zeta_R(2s - p + 1)$$

$$+ 4\pi^s \left(a_{11} - \vec{a}_{p-1}^T A_{p-1}^{-1} \vec{a}_{p-1}\right)^{(p-1)/4 - s/2}$$

$$\cdot \sum_{n=1}^{\infty} \sum_{\vec{m} \in \mathbb{Z}_{1/2}^{p-1}}{}' \cos\left(2\pi \vec{m}^T A_{p-1}^{-1} \vec{a}_{p-1} n\right) n^{(p-1)/2 - s}$$

$$\cdot \left(\vec{m}^T A_{p-1}^{-1} \vec{m}\right)^{s/2 - (p-1)/4}$$

$$\left. \cdot K_{(p-1)/2 - s}\left[2\pi n \sqrt{\left(a_{11} - \vec{a}_{p-1}^T A_{p-1}^{-1} \vec{a}_{p-1}\right) \vec{m}^T A_{p-1}^{-1} \vec{m}}\right] \right\}. \quad (4.41)$$

This is also a recurrent expression, an alternative to (4.38), obtained with the help of a different strategy.

Remarkably enough, it is easy to resolve this recurrence explicitly, and indeed to obtain a *closed formula* for this case (we shall write the dimensions of the submatrices of A as subindices). The result being

$$\zeta_{A_p}(s) \equiv \zeta_{A_p,\vec{0},0}(s)$$

$$= \frac{2^{1+s}}{\Gamma(s)} \sum_{j=1}^{p} (\det A_{p-j})^{-1/2} \left\{ \pi^{(p-j)/2} \left(a_{jj} - \vec{a}_{p-j}^T A_{p-j}^{-1} \vec{a}_{p-j} \right)^{(p-j)/2-s} \right.$$

$$\cdot \Gamma\left(s - (p-j)/2\right) \zeta_R(2s - p + j)$$

$$+ 4\pi^s \left(a_{jj} - \vec{a}_{p-j}^T A_{p-j}^{-1} \vec{a}_{p-j} \right)^{(p-j)/4-s/2}$$

$$\cdot \sum_{n=1}^{\infty} \sum_{\vec{m}_{p-j} \in \mathbb{Z}_{1/2}^{p-j}} {}' \cos\left(2\pi \vec{m}_{p-j}^T A_{p-j}^{-1} \vec{a}_{p-j} n\right) n^{(p-j)/2-s}$$

$$\cdot \left(\vec{m}_{p-j}^T A_{p-j}^{-1} \vec{m}_{p-j} \right)^{s/2-(p-j)/4}$$

$$\left. \cdot K_{(p-j)/2-s}\left[2\pi n \sqrt{\left(a_{jj} - \vec{a}_{p-j}^T A_{p-j}^{-1} \vec{a}_{p-j} \right) \vec{m}_{p-j}^T A_{p-j}^{-1} \vec{m}_{p-j}} \right] \right\}. \quad (4.42)$$

With a similar notation as above, here A_{p-j} is the submatrix of A_p composed of the last $p - j$ rows and columns. Moreover, a_{jj} is the jth diagonal component of A_p, while $\vec{a}_{p-j} = (a_{jj+1}, \ldots, a_{jp})^T = (a_{j+1j}, \ldots, a_{pj})^T$, and $\vec{m}_{p-j} = (n_{j+1}, \ldots, n_p)^T$. Again, this is an extension of the Chowla–Selberg formula to the case in question. It exhibits all the same good properties. Physically, it corresponds to the homogeneous, massless case. It is to be viewed, in fact, as *the* genuine multidimensional extension of the Chowla–Selberg formula. We shall call it ECS3.

Diagonal Subcase Let us particularize once more to the diagonal case, with $\vec{c} = \vec{0}$, which is quite important in practice and gives rise to more simple expressions. For the recurrence, we have

$$\zeta_{A_p}(s) = \zeta_{A_{p-1}}(s) + \frac{2^{1+s}}{\Gamma(s)} (\det A_{p-1})^{-1/2} \left[\pi^{(p-1)/2} a_1^{(p-1)/2-s} \Gamma\left(s - (p-1)/2\right) \right.$$

$$\cdot \zeta_R(2s - p + 1)$$

$$+ 4\pi^s a_1^{(p-1)/4-s/2} \sum_{n=1}^{\infty} \sum_{\vec{m} \in \mathbb{Z}^{p-1}} {}' n^{(p-1)/2-s} \left(\vec{m}^T A_{p-1}^{-1} \vec{m} \right)^{s/2-(p-1)/4}$$

$$\left. \cdot K_{(p-1)/2-s}\left(2\pi n \sqrt{a_1 \vec{m}^T A_{p-1}^{-1} \vec{m}} \right) \right]. \quad (4.43)$$

As above, we can solve this finite recurrence and obtain the following simple and explicit formula for this case:

$$
\zeta_{A_p}(s) = \frac{2^{1+s}}{\Gamma(s)} \sum_{j=0}^{p-1} (\det A_j)^{-1/2} \Bigg[\pi^{j/2} a_{p-j}^{j/2-s} \Gamma(s-j/2) \zeta_R(2s-j)
$$

$$
+ 4\pi^s a_{p-j}^{j/4-s/2} \sum_{n=1}^{\infty} \sum_{\vec{m}_j \in \mathbb{Z}^j}{}' n^{j/2-s} \left(\vec{m}_j^t A_j^{-1} \vec{m}_j \right)^{s/2-j/4}
$$

$$
\cdot K_{j/2-s} \left(2\pi n \sqrt{a_{p-j} \vec{m}_j^t A_j^{-1} \vec{m}_j} \right) \Bigg],
\tag{4.44}
$$

with $A_p = \mathrm{diag}(a_1, \ldots, a_p)$, $A_j = \mathrm{diag}(a_{p-j+1}, \ldots, a_p)$, $\vec{m}_j = (n_{p-j+1}, \ldots, n_p)^T$, and ζ_R the Riemann zeta function. Note again the fact that this and (4.42) are *explicit* expression for the multidimensional, generalized Chowla–Selberg formula and, in this way, they go beyond any result obtained previously. We name this formula ECS3d.

It is immediate to see that the term for $j = 0$ in the sum yields the last term, $\zeta_{A_1}(s)$, of the recurrence, that is:

$$
\zeta_{A_1}(s) = \sum_{n_p=-\infty}^{+\infty}{}' \left(\frac{a_p}{2} n_p^2 \right)^{-s} = 2^{1+s} a_p^{-s} \zeta_R(2s).
\tag{4.45}
$$

It exhibits a pole, at $s = 1/2$ which is spurious—it is actually *not* a pole of the whole function (since it cancels, in fact, with another one coming from the next term, with further cancellations of this kind going on, each term with the next). Concerning the pole structure of the resulting zeta function, as given by (4.44), it is not difficult to see that *only* the pole at $s = p/2$ is actually there (as it should). It is in the last term, $j = p - 1$, of the sum, and it has the correct residue, namely

$$
\mathrm{Res}\, \zeta_{A_p}(s) \big|_{s=p/2} = \frac{(2\pi)^{p/2}}{\Gamma(p/2)} (\det A_p)^{-1/2}.
\tag{4.46}
$$

The rest of the seem-to-be poles at $s = (p - j)/2$ are not such: they compensate among themselves, one term of the sum with the next, adding pairwise to zero.

Summing up, this formula, (4.44), provides a convenient analytic continuation of the zeta function to the whole complex plane, with its only simple pole showing up explicitly. Aside from this, the finite part of the first sum in the expression is quite easy to obtain, and the remainder—an awfully looking multiple series—is in fact an extremely-fast convergent sum, yielding a small contribution, as happens in the CS formula. In fact, since it corresponds to the case $q = 0$, this expression should be viewed as *the* extension of the original Chowla–Selberg formula—for the zeta function associated with an homogeneous quadratic form in two dimensions—to an arbitrary number, p, of dimensions. The rest of the formulas above provide also extensions of the original CS expression.

The general case of a quadratic+linear+constant form has been here thus completed. As we clearly see, the main formulas corresponding to the three different subcases, namely ECS1 (4.36), ECS2 (4.39), and ECS3 (4.42), are in fact quite distinct and one cannot directly go from one to another by adjusting some parameters.

For the sake of completeness, we must mention the following. Notice that all cases considered here correspond to having a non-identically-zero quadratic form Q. For Q identically zero, that is, the linear+constant (or affine) case, the formulas for the analytic continuation are again quite different from the ones above. The corresponding zeta function is called Barnes' zeta function. This case has been thoroughly studied in Ref. [125].

4.3 Numerical Analysis of the Inhomogeneous Generalized Epstein–Hurwitz Zeta Function

The inhomogeneous generalized (Epstein–Hurwitz like) multi-dimensional series

$$E_m\left(s; a_1, \ldots, a_m; c_1, \ldots, c_m; c^2\right)$$

$$\equiv \sum_{n_1,\ldots,n_m=0}^{\infty} \left[a_1(n_1 + c_1)^2 + \cdots + a_m(n_m + c_m)^2 + c^2\right]^{-s} \quad (4.47)$$

can be reduced (as we have already seen above), by means of a non-trivial asymptotic recurrence, to the one-dimensional case

$$E_1\left(s; a, b^2\right) = \sum_{n=0}^{\infty} \left[(n + a)^2 + b^2\right]^{-s}, \quad (4.48)$$

which we will study here in full detail. In particular, asymptotic expansions for F and its derivatives $\partial F/\partial s$ and $\partial F/\partial a$—together with analytical continuations of the same in the variable s—will be explicitly obtained using zeta-function techniques. Several plots and tables of the numerical results will be given. In Chap. 6, some explicit applications of these expressions to the regularization, by means of Hurwitz zeta-functions, of different problems which have appeared recently in the physical literature, will be investigated.

As we have seen above, when complicated theories over spacetimes with topologies of increasing non-triviality are considered (relevant, e.g., for the description of the early stages of our universe), the results of the regularization are expressed in terms of rather non-elementary zeta functions, such as Epstein's one (for the case of the torus compactification), Epstein–Hurwitz's, and generalizations thereof. In other words, one has to deal with complicated Dirichlet series, sometimes well known to mathematicians (who make good use of those in number theory) but sometimes unknown to them, and therefore deserving careful analysis.

Here we shall investigate the case of multiple series of the generalized Epstein–Hurwitz type (that we will call here F)

$$F\left(s; a, b^2\right) \equiv E_1\left(s; a; b^2\right) = \sum_{n=0}^{\infty}\left[(n+a)^2 + b^2\right]^{-s}, \qquad (4.49)$$

and their extension in the form of multidimensional series (4.47). This kind of zeta functions appear often in different applications of quantum physics where regularization techniques are needed, in particular, when one deals with a massive quantum field theory in a (totally or partially) compactified spacetime—spherical or toroidal compactification, for instance. Aside from the interest that a detailed mathematical study of these functions may have on its own (e.g. in number theory), what is actually needed for most physical applications—as we shall explicitly see later—is always the *numerical* value of the analytical continuation of such expressions, and of their *derivatives* with respect to the variable s and to the parameter a, for negative (half)integer values of s and for a few simple fractional values of a (like $a = 1/2, 1/4, 3/4, \ldots$). Concerning the parameter b^2, usually we need the large- or small-b^2 behavior of these series only.

The final results involve analytical continuation of the functions and also a principal part prescription in order to deal with (possible) poles. This procedure has already been checked in several situations, see [42, 126]. The results will be always given in terms of sums of Hurwitz zeta functions and generally under the form of asymptotic expansions.

4.3.1 Asymptotic Expansions of the Function and Its Derivatives with Respect to the Variable and Parameters

Coming back to our function of generalized Epstein–Hurwitz type $F(s; a, b^2)$, (4.49), and as a particular case of the above analysis, two complementary asymptotic expansions, for large and for small b^2, respectively, are not difficult to obtain. The first is

$$F\left(s; a, b^2\right) \sim \frac{b^{-2s}}{\Gamma(s)} \sum_{m=0}^{\infty} \frac{(-1)^m \Gamma(m+s)}{m!} b^{-2m} \zeta(-2m, a)$$

$$+ \frac{\sqrt{\pi}\,\Gamma(s - 1/2)}{2\Gamma(s)} b^{-2s+1}$$

$$+ \frac{2\pi^s b^{-s+1/2}}{\Gamma(s)} \sum_{n=1}^{\infty} n^{s-1/2} \cos(2\pi na)\, K_{s-1/2}(2\pi nb), \quad (4.50)$$

which is an asymptotic expansion valid in principle for large b^2. However, a numerical investigation has shown that its range of validity is rather wide, so that for $b^2 \simeq 0.5$ or so it still gives quite acceptable results.

Table 1 Numerical values of the quotient of the last term Re $F(s; a, b^2)$ of the asymptotic expansion of $F(s; a, b^2)$ by the whole function, for $a = 1/4$ and several values of b^2. The resulting intervals correspond to the extreme values of the quotient in the range $s \in [-5, 2]$, which covers by far all cases which appear in practice

b^2	Re $F(s; 1/4, b^2)/F(s; 1/4, b^2)$
10	10^{-21}–10^{-22}
1.5	10^{-8}–10^{-9}
1	10^{-7}–10^{-8}
0.5	10^{-6}–10^{-7}
0.1	10^{-7}–10^{-10}
0.01	10^{-11}–10^{-18}

The last term of this asymptotic expansion (4.50) can actually be suppressed, as is clear from the definition of asymptoticity itself. In fact, to be sure of that, in Table 1 we present the results of a numerical evaluation of the quotient of the last term, Re $F(s; a, b^2)$ by the whole function $F(s; a, b^2)$, for $a = 1/4$ and several values of b^2, and for a wide range for s, which covers all cases that appear in practice (namely, $-5 \leq s \leq 2$).

We see from Table 1 that, with great precision, the last term can be suppressed from (4.50), in other words that asymptoticity is fulfilled, to all purposes, for any practical use of the formula as will be described below. We are thus left with the much simpler form of the expansion (4.50)

$$F\left(s; a, b^2\right) \sim \frac{b^{-2s}}{\Gamma(s)} \sum_{m=0}^{\infty} \frac{(-1)^m \Gamma(m+s)}{m!} b^{-2m} \zeta(-2m, a)$$

$$+ \frac{\sqrt{\pi}\,\Gamma(s - 1/2)}{2\Gamma(s)} b^{-2s+1}. \tag{4.51}$$

This series usually has its optimal truncation around the 10th term. A detailed numerical study has shown that, in fact, for the range of values considered here the optimal truncation is obtained with exactly ten terms (the eleventh term provides the absolute minimum), being the order of the error bound of 10^{-4}. Actually in this analysis we have also taken into account the first term of the exponentially convergent series, that we here just dismiss in order to simplify the expressions. The contribution of this term is certainly less than the error bound in the whole range considered. Similar analysis carried out with the functions M_2 and S_α lead to the same conclusions in the most general case, just with a possible increase of the error bound, that may reach 10^{-2} at worst (always for the region of parameters here considered).[5]

---

[5]I am indebted with S. Rafels-Hildebrandt for this additional analysis.

Table 2 A comparison of the numerical values which come from the two expressions corresponding to the function $F(s; a, b^2)$ for $a = 1/4$ and a choice of values of the parameter b^2 and of the variable s. We see that the coincidence for values of b^2 bigger than one and s negative is quite remarkable. For small b^2, say $b^2 < 2$, the second expression is the one to be used, while for bigger b^2's the first is more convenient

s	b^2	$F1(s; 1/4, b^2)$	$F2(s; 1/4, b^2)$
-1	1	0.4675	0.5547
-1.7	1	-0.5353	-0.5507
-2.1	1	0.3889	0.3929
-3.1	1	0.3621	0.3598
-4.1	1	0.3351	0.2541
-1.7	1.5	-1.4302	-1.4311
-1.7	2	-2.7999	-2.8000
-1.7	4	-13.8399	-13.8399
-1.7	10	-110.798	-110.798
-1.7	20	-525.568	-525.566
-3.7	20	-152203	-152202.9
1.7	20	0.02614	0.025644

An alternative expansion for the function (4.49) is the following

$$F\left(s; a, b^2\right) = \sum_{n=0}^{[b^2]} \left[(n + a)^2 + b^2\right]^{-s}$$

$$\cdot \sum_{m=0}^{\infty} \frac{(-1)^m \Gamma(m + s) b^{2m}}{m! \Gamma(s)} \left[\zeta(2m + 2s, a) - \sum_{n=0}^{[b^2]} (n + a)^{-2m-2s}\right],$$

$$(4.52)$$

which is valid for finite b^2. It is obtained using binomial expansion and, in particular, absolute convergence of the binomial series for values of the argument < 1 (for a detailed explanation, see [126]). Both these expansions are plotted in Figs. 3 and 4, for some convenient values of a and b^2 (such as $a = 1/2, 1/4, b = 0.1, 1.5, 10$, the ones most employed in physical applications) and for a range of s which also covers all practical cases. In those figures, we denote by $F1$ the asymptotic expression for F given by (4.50), and by $F2$ the expression given by (4.52). A careful numerical comparison of the two methods—leading namely to (4.50) and (4.52)—shows that both give very similar results for values of b^2 which are finite and bigger than one. In this case the best thing to do in practice is to use (4.51) and its derivatives, which is simple enough and provides a good approximation. On the other hand, when $b^2 < 1$ then that formula ceases to be accurate and the binomial expansion (4.52)—which is then reduced precisely to its most simple possible form—is the one to be employed. A choice of numerical results, for comparison, is given in Table 2.

However, in more involved applications (see [126, 127]) different partial derivatives of the function $F(s; a, b^2)$ (basically, the first derivatives with respect to s and a) need to be calculated.

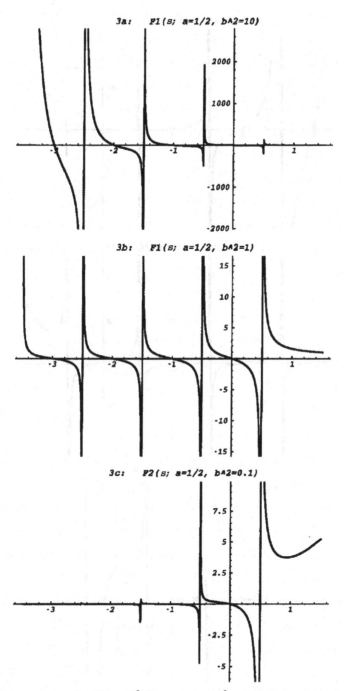

Fig. 3 (a) Plot of the function $F1(s; a, b^2)$ for $a = 1/2$ and $b^2 = 10$, in the interval $s \in [-3.5, 1.5]$. (b) Plot of the function $F1(s; a, b^2)$ for $a = 1/2$ and $b^2 = 1$, in the interval $s \in [-3.5, 1.5]$. (c) Plot of the function $F2(s; a, b^2)$ for $a = 1/2$ and $b^2 = 0.1$, in the interval $s \in [-3.5, 1.5]$

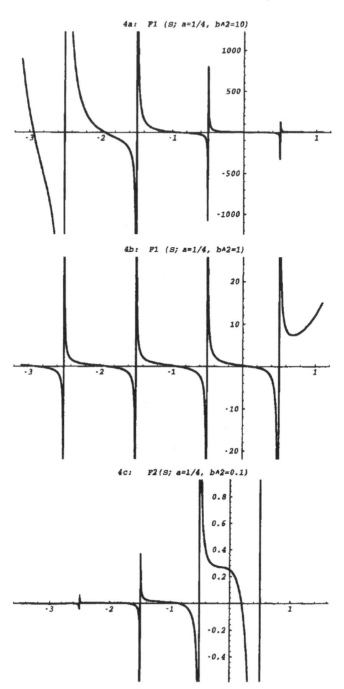

Fig. 4 (**a**) Plot of the function $F1(s; a, b^2)$ for $a = 1/4$ and $b^2 = 10$, in the interval $s \in [-3.2, 1.2]$.
(**b**) Plot of the function $F1(s; a, b^2)$ for $a = 1/4$ and $b^2 = 1$, in the interval $s \in [-3.2, 1.2]$. (**c**) Plot
of the function $F2(s; a, b^2)$ for $a = 1/4$ and $b^2 = 0.1$, in the interval $s \in [-3.5, 1.5]$

The a-derivative is not difficult to obtain

$$\frac{\partial}{\partial a} F\left(s; a, b^2\right) \simeq \sum_{n=0}^{\infty} \frac{(-1)^n \Gamma(s+n)}{n! \Gamma(s)} b^{-2s-2n} \tilde{\zeta}(-2n, a)$$

$$- \frac{4\pi^{s+1} b^{-s+1/2}}{\Gamma(s)} \sum_{n=1}^{\infty} n^{s+1/2} \sin(2\pi n a) K_{s-1/2}(2\pi n b),$$

$$(4.53)$$

where

$$\tilde{\zeta}(z, a) \equiv \frac{\partial}{\partial a} \zeta(z, a), \tag{4.54}$$

and is the asymptotic expansion for the a-derivative of F in the case $b \geq 1$. For the derivative of the Hurwitz zeta function, the following simple relation holds

$$\tilde{\zeta}(z, a) = -z\zeta(z+1, a). \tag{4.55}$$

In particular, we have that $\tilde{\zeta}(0, 1/2) = -1$, more generally:

$$\tilde{\zeta}(0, a) = \frac{\partial}{\partial a} \zeta(0, a) = \frac{\partial}{\partial a}\left(\frac{1}{2} - a\right) = -1. \tag{4.56}$$

The same result is obtained from (4.55) in the limit $s \to 0$. On the other hand, using that

$$\frac{\Gamma(n-k)}{\Gamma(-k)} = (-1)^n \frac{k!}{(k-m)!}, \tag{4.57}$$

for $n \leq k$, and that this is zero for $n \geq k+1$, the formula for the derivative at negative integer values of the variable s can be simplified. Putting everything together, we get

$$\frac{\partial}{\partial a} F\left(-k; a, b^2\right) \sim 2 \sum_{n=1}^{k} \frac{k! b^{2k-2n}}{(n-1)!(k-n)!} \zeta(1-2n, a), \tag{4.58}$$

for $k \neq 0$, and it is equal to -1 for $k = 0$.

For the case of small b^2, the corresponding asymptotic expansion is easier to obtain from the alternative expression (4.52) (for explicit uses of this formula, see [126]). Its partial derivative with respect to a is given by

$$\frac{\partial}{\partial a} F\left(s; a, b^2\right) \sim -2s \sum_{n=0}^{[b^2]} \left[(n+a)^2 + b^2\right]^{-s-1} (n+a)$$

$$+ 2 \sum_{m=0}^{\infty} \frac{(-1)^{m+1} \Gamma(m+s+1) b^{2m}}{m! \Gamma(s)}$$

Table 3 Numerical values for the derivative of $F(s; a, b^2)$ with respect to a at a sample of points. The first three columns are the results of the calculation with the first of the two series expansions for F, while the last column has been obtained by using the second expansion

k	$F_a'(-k; 1/4, 10)$	$F_a'(-k; 1/4, 3/2)$	$F_a'(-k; 1/4, 1)$	$F_a'(-k; 1/4, 0.1)$
0	-1	-1	-1	-1
1	0.02083	0.02083	0.02083	-0.079167
2	0.4148	0.0607	0.0398	-0.007656
3	6.1957	0.1328	0.0574	-0.000561
4	82.254	0.2587	0.0737	-0.000111
5	1023.79	0.4730	0.0890	0.000027

$$\cdot \left[\zeta(2m + 2s + 1, a) - \sum_{n=0}^{[b^2]} (n + a)^{-2m-2s-1} \right]. \quad (4.59)$$

In the particular case $s = -k$, $k = 0, 1, 2, \ldots$, using

$$\frac{\zeta(1 + 2\epsilon, a)}{\Gamma(-k + \epsilon)} = \frac{1}{2}(-1)^k k! + \left(\frac{1}{2a} - \ln a \right) \epsilon + \mathcal{O}(\epsilon^2), \quad (4.60)$$

we obtain

$$\frac{\partial}{\partial a} F\left(-k; a, b^2 \right)$$

$$\sim -b^{2k} + 2k \sum_{n=0}^{[b^2]} \left[(n + a)^2 + b^2 \right]^{k-1} (n + a)$$

$$+ 2 \sum_{m=0}^{k-1} \frac{k! b^{2m}}{m!(k - m - 1)!} \left[\zeta(2m - 2k + 1, a) \right.$$

$$\left. - \sum_{n=0}^{[b^2]} (n + a)^{-2m+2k-1} \right]. \quad (4.61)$$

Table 3 shows the value of the derivative of $F(s; a, b^2)$ with respect to a, for several values of the parameters, which cover the cases that usually appear in practical applications.

Concerning the s-derivative, a direct calculation, which makes use of the reflection formulas (and can be justified with the same arguments of [98]) yields:

$$\frac{\partial}{\partial s} F\left(s; a, b^2 \right) \bigg|_{s=-k}$$

$$\sim b^{2k}\sum_{m=0}^{k}\frac{k!b^{-2m}}{m!(k-m)!}\zeta(-2m,a)$$

$$\cdot\left(\frac{1}{k}+\frac{1}{k-1}+\cdots+\frac{1}{k-m+1}-\ln b^2\right)$$

$$+b^{2k}\sum_{m=k+1}^{\infty}\frac{(-1)^{m+k}k!(m-k-1)!}{m!}b^{-2m}\zeta(-2m,a)$$

$$+\frac{\sqrt{\pi}}{2}(-1)^k k!\Gamma(-k-1/2)b^{2k+1}$$

$$+2(-\pi)^{-k}k!b^{k+1/2}\sum_{n=1}^{\infty}n^{-k-1/2}\cos(2\pi na)K_{-k-1/2}(2\pi nb).$$

$$(4.62)$$

Alternatively, we can use the second procedure as described above, (4.52), which results in

$$\frac{\partial}{\partial s}F\left(s;a,b^2\right)\big|_{s=-k}$$

$$\sim-\sum_{n=0}^{[b^2]}\left[(n+a)^2+b^2\right]^k\ln\left[(n+a)^2+b^2\right]$$

$$+2\sum_{n=0}^{[b^2]}(n+a)^{2k}\ln(n+a)$$

$$+2\zeta'(-2k,a)+\sum_{m=1}^{k}\frac{k!b^{2m}}{m!(k-m)!}\left(\frac{1}{k}+\frac{1}{k-1}+\cdots+\frac{1}{k-m+1}\right)$$

$$\cdot\left[\zeta(2m-2k,a)-\sum_{n=0}^{[b^2]}(n+a)^{-2m+2k}\right]$$

$$+\sum_{m=1}^{k}\frac{k!b^{2m}}{m!(k-m)!}\left[2\zeta'(2m-2k,a)+2\sum_{n=0}^{[b^2]}(n+a)^{-2m+2k}\ln(n+a)\right]$$

$$+\sum_{m=k+1}^{\infty}\frac{(-1)^{m+k}k!(m-k-1)!}{m!}b^{2m}$$

$$\cdot\left[\zeta(2m-2k,a)-\sum_{n=0}^{[b^2]}(n+a)^{-2m+2k}\right],$$

$$(4.63)$$

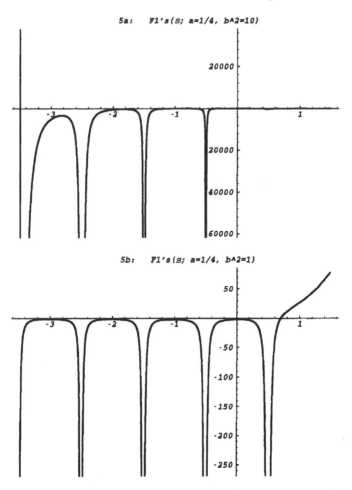

Fig. 5 (**a**) Plot of the s-derivative of the function $F1(s; a, b^2)$ for $a = 1/4$ and $b^2 = 10$, in the interval $s \in [-3.5, 1.5]$. (**b**) Plot of the s-derivative of the function $F1(s; a, b^2)$ for $a = 1/4$ and $b^2 = 1$, in the interval $s \in [-3.5, 1.5]$

where, again, we have restricted ourselves to non-positive integer values of the argument, $s = -k, k = 0, 1, 2, 3, \ldots$. Plots of the s and a derivatives of F for $a = 1/4$ and several values of b^2 are represented in Figs. 5 and 6.

On the other hand, numerical values for the derivative of $F(s; a, b^2)$ with respect to s at several selected points are given in Table 4. The numbers are quite reliable, in fact, we have checked that use of the explicit expression (4.62) for the partial derivative differs from direct numerical computation of the derivative by less than one per mil.

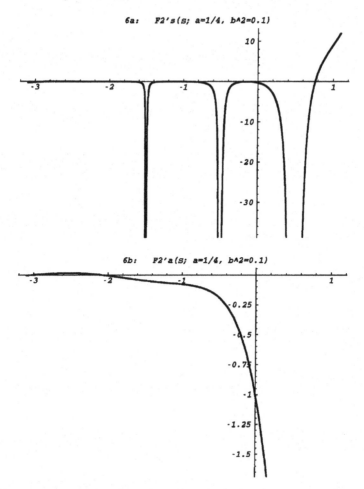

Fig. 6 (a) Plot of the s-derivative of the function $F2(s; a, b^2)$ for $a = 1/4$ and $b^2 = 0.1$, in the interval $s \in [-3.1, 1.1]$. (b) Plot of the a-derivative of the function $F2(s; a, b^2)$ for $a = 1/4$ and $b^2 = 0.1$, in the interval $s \in [-3.1, 0.2]$

Table 4 Numerical values of the derivative of $F(s; a, b^2)$ with respect to s at several selected points. The first three columns have been calculated by means of the first expression for F, and the last one by using the second

k	$F_s'(-k; 1/4, 10)$	$F_s'(-k; 1/4, 3/2)$	$F_s'(-k; 1/4, 1)$	$F_s'(-k; 1/4, 0.1)$
0	−10.509	−3.904	−1.972	1.05664
1	−71.936	−3.986	−2.2303	−0.03363
2	−586.55	−4.808	−1.621	−0.02111
3	−5105.3	−6.223	−1.451	−0.00035
4	−45974	−8.330	−1.249	−0.00776
5	−422725	−11.411	−1.179	0.02029

Chapter 5
Physical Application: The Casimir Effect

Although some examples and applications of the formulas for analytical continuation have already been given in the preceding chapters, in the present one we shall properly start with the discussion in length of the applications of the explicit zeta function regularization method. In this sense, the Casimir effect is introduced, along with the Casimir–Polder version, and its relation with the van der Waals forces, the London theory, and the generalization of the Casimir setup in terms of the very far reaching Lifshitz theory are discussed. Also the "mistery" of the Casimir effect, its local formulation and the definition of the Casimir energy in terms of the fluctuations of the quantum vacuum. All the cases we are going to consider here will correspond precisely to the situation to which we have restricted ourselves in this work, namely the case when the spectrum of the Hamiltonian of our physical system is known explicitly. Already this startpoint, that might be considered as rather particular, gives rise to quite interesting situations from the physical point of view (and from the mathematical one, too), that are addressed in this chapter.

5.1 Essentials of the Casimir Effect

5.1.1 The Original Casimir Effect

The Casimir effect takes its name after the Dutch prominent physicist H.B.G. Casimir, who in 1948 published a paper in the Proceedings of the Academy of Sciences of the Netherlands where a rather remarkable property, namely the attraction of two neutral metallic plates, was predicted theoretically [128, 129]. In all the research papers and reviews about the Casimir effect that have been published in the last years [49, 130, 131], this paper by Casimir is taken as the indubious beginning of a whole branch of research, which aims nowadays at answering profound questions about the vacuum structure of quantum field theory. However, it is difficult to get a clear idea from these—on the other hand excellent review papers—of what the contribution of Casimir was precisely, or of what was the specific physical context in which his paper appeared. Some authors go even further and attribute to Casimir deep ideas about quantum field theory that by no means could he have had at this

E. Elizalde, *Ten Physical Applications of Spectral Zeta Functions*,
Lecture Notes in Physics 855,
DOI 10.1007/978-3-642-29405-1_5, © Springer-Verlag Berlin Heidelberg 2012

time. The interest of the subject, which is reflected by the increasing number of papers which are dealing with it, deserves a proper clarification of several points.

To start with, it is fair to say that the 1948 paper by Casimir attracted comparatively small attention during the following two or three decades. For instance, another paper by Casimir and Polder [132], which was published in Physical Review also in 1948, got by far much more citations from experimental and theoretical colleagues. This second paper is nowadays considered a mere addition or an initial stimulus to the first one, viewed now as the fundamental paper about the Casimir effect. Maybe part of this puzzle can be explained by the importance and availability of the Physical Review (in comparison with the aforementioned Dutch journal). However, even in contributions where the two papers are mentioned, the one by Casimir alone deserves no special comment, i.e. in no way was it singularized with respect to the other one coauthored by Polder.

5.1.2 Connection with the van der Waals Forces and the London Theory

A second point to be remarked is the following [133]. Nowadays, when dealing with the Casimir effect itself, a particular emphasis is usually put in the spectacularity of the same, that is, in the fact that two non-charged plates do attract themselves in the vacuum. One needs to understand that this is actually much more mysterious today that it was in 1948. At that time, 67 years after the publication of the work of D. van der Waals [134]—where his famous weak attractive forces between neutral molecules were introduced—and already 18 years after de formulation by F. London of his celebrated theory [135]—which gave a precise (for that time) explanation of the nature and strength of the van der Waals forces as due to the interaction of the fluctuating electric dipole moments of the neutral molecules—there was nothing specially mysterious about two neutral bodies attracting each other. Van der Waals forces play a very important role in biology and in the medical sciences. They are particularly significant in surface phenomena, such as adhesion, colloidal stability and foam formation. One could dare to say that they are the most important physical forces controlling living beings and life processes. Three different classes of van der Waals forces can in principle be distinguished: orientation, induction, and dispersion forces [130]. The ones involved in the attraction of the plates correspond in this classification to the third group.

5.1.3 The Specific Contribution of Casimir and Polder: Retarded van der Waals Forces

The works of Casimir and Casimir and Polder, addressed rather an (of course important but) more technical point: the fact that the polarization of the neighboring molecules (or atoms) induced by a given molecule (atom) is delayed as a consequence of the finiteness of the velocity of light. So these forces could be termed as long-range retarded dispersion van der Waals forces. This was clearly noticed in the experiments by the fact that when the molecules were separated far enough (always in the range of the microns, so that the effect could make any sense), the power law corresponding to the attractive force between two given molecules changed to the inverse eighth power,

$$F = \frac{C_2}{r^8} \tag{5.1}$$

(C_2 being a constant and r the distance between the two molecules), compared with the inverse seventh power obtained in London's theory:

$$F = \frac{C_1}{r^7}, \tag{5.2}$$

typical of the van der Waals forces for very close molecules (which did not feel so much the retardation effect due to the finite velocity of interaction). In particular, it has become famous the expression obtained by Casimir and Polder for the potential energy U corresponding to two atoms separated by a distance r and whose static polarizabilities are α_1 and α_2, respectively:

$$U = -\frac{23\hbar c}{4\pi} \frac{\alpha_1 \alpha_2}{r^7}. \tag{5.3}$$

It is now (and was then) a matter of an elementary exercise on surface integration to obtain the force per unit surface (that is, the pressure) which attracts two neutral, parallel, metallic, perfectly conducting plates of infinite extension in the vacuum, under the hypothesis that they are formed by a rarefied distribution of neutral, polarizable atoms. However, Casimir used a novel technique: that of calculating the effect due to the zero-point energy of the electromagnetic field. Simple dimensional reasons show immediately that the just mentioned power laws give rise, respectively, to an inverse third power, if the plates are very close (distance say less or equal than 0.01 micron, i.e. 100 Å, which is the penetration depth of electromagnetic waves in the metal),

$$P = \frac{A}{6\pi d^3}, \tag{5.4}$$

and to an inverse fourth power, if they are more separated (say somewhat above 0.02 microns)

$$P = \frac{B}{d^4}. \tag{5.5}$$

This last case was the one explicitly calculated by Casimir in his seminal paper [128, 129] with the result:

$$P = \frac{0.013}{d^4} \frac{\text{dynes}}{\text{cm}^2}, \tag{5.6}$$

where the distance in this expression has to be given in microns, the basic unit length for this kind of calculations, as mentioned already. The calculations can be easily extended to different geometrical configurations. For instance, for a hemisphere of radius R held at distance d from an infinite plane, the attractive force when they are very close was given by London's theory as

$$F = \frac{AR}{6d^2}. \tag{5.7}$$

On the other hand, when they are a bit more separated, the retarded interaction changed this result to

$$F = \frac{2\pi BR}{3d^3}. \tag{5.8}$$

The value obtained by Casimir and Polder for the potential energy corresponding to a particle of electric polarizability α, inside a cavity of a perfectly conducting material and separated a distance r from the flat wall was

$$U = -\frac{3\hbar c}{8\pi} \frac{\alpha}{r^4}. \tag{5.9}$$

5.1.4 The Lifshitz Theory

E.M. Lifshitz, in a no less important paper [136] than those previously referred to (submitted to the Russian JETP in 1954 and whose English translation was published in 1956), developed a different theory in order to deal with the two major difficulties of London's theory, namely the already mentioned one, that it did not take into account the finite velocity of propagation of the electromagnetic interaction (this had been already taken care of by Casimir and Polder, see also the generalization of Ref. [137]), and a second one, namely the fact that the van der Waals force is not additive. This prevents to treat the problem of extensive bodies in a proper way as composed of elementary constituents (atoms or molecules), and to derive the force between macroscopic bodies by integration of the forces which exist between the elementary constituents—unless one makes the hypothesis (advanced before) of considering a very dilute distribution of constituents in the extensive bodies, but this is an unrealistic assumption.

Lifshitz's theory started from the opposite direction, treating matter as a continuum with a well-defined, frequency-dependent dielectric susceptibility. It was a completely closed theory: it could deal with any kind of material bodies, it explained in a precise and continuous way the transition from one power-law to the other (due to the retardation effect) when the distance is increased, and it contained the formulas of London for the elementary constituents of matter, and of Casimir and Polder for the perfectly conducting, neutral metallic plates, as limiting and particular cases, respectively, what was proven by Dzyaloshinskii, Lifshitz and Pitaevskii in 1961 [138]. There the picture of the interacting bodies in Lifshitz's theory was that of two media filling half-spaces with plane parallel boundaries separated from one another by a certain distance d. Just as for the case of the random force introduced in the theory of Brownian motion, a 'random' field was introduced into the Maxwell equations of motion. In the case of large (that is, not so small) separations, Lifshitz's formula for the force per unit area between two parallel plates separated a distance d one from the other reduced to the following. For plates of an infinitely conducting metal:

$$P = \frac{\hbar c \pi^2}{240 d^4};\tag{5.10}$$

for two identical dielectrics of dielectric constant ε_0:

$$P = \frac{\hbar c \pi^2}{240 d^4} \left(\frac{\varepsilon_0 - 1}{\varepsilon_0 + 1} \right)^2 \varphi(\varepsilon_0),\tag{5.11}$$

where $\varphi(\varepsilon_0)$ is a function defined by the theory and which has the following behavior for $\varepsilon_0 \to \infty$:

$$\varphi(\varepsilon_0) \simeq 1 - \frac{1.11}{\sqrt{\varepsilon_0}} \ln \frac{\varepsilon_0}{7.6};\tag{5.12}$$

for a metal (infinitely conducting, $\varepsilon = \infty$) and a dielectric of constant ε_0:

$$P = \frac{\hbar c \pi^2}{240 d^4} \frac{\varepsilon_0 - 1}{\varepsilon_0 + 1} \varphi(\varepsilon_0);\tag{5.13}$$

also, for two individual atoms belonging to materials of dielectric constants ε_{10} and ε_{20}, respectively, Lifshitz obtained the attractive force between them as a limiting case of the formula for continuous media, assuming that both media were sufficiently rarefied:

$$F = \frac{23 \hbar c}{640 \pi^2 d^4} (\varepsilon_{10} - 1)(\varepsilon_{20} - 1),\tag{5.14}$$

from which the preceding formula of Casimir and Polder por the potential energy follows immediately.

A most important point of this theory was the fact that the general equations derived by Lifshitz to calculate the dispersion force requires only information about the dielectric properties of the bodies (in particular, the dielectric susceptibility of the bodies as a function of the frequency), and this information can, in principle, be obtained from independent spectroscopic measurements. Thus Lifshitz's theory could be applied by Parsegian and Ninham (in 1970) to a detailed study of the dispersion forces between biological membranes [139–141]. A very original contribution of the theory was also to consider the effect of temperature on the force of interaction. However, this part of Lifshitz's theory disagreed with subsequent (independent) calculations by Sauer [142], Mehra [143], and Brown and MacLay [41], who agreed among themselves and said that Lifshitz's results concerning this point were in error.

To make the story short, starting with Lamoreaux's experiment [144] this field has undergone a complete revolution and this is today considered to be the first actual verification of the Casimir effect. The much higher accuracy of the new experiments with respect to the older ones (described in the first edition) seem to make these last irrelevant now. Updated sources of information are the website of the ESF CASIMIR project and James F. Babb's webpage [145].

5.2 The Casimir Effect in Quantum Field Theory

5.2.1 The Local Formulation of the Casimir Effect

The paper by L.S. Brown and G.J. MacLay, published 21 years after the work of Casimir, was specially significant from the theoretical point of view, a kind of milestone in the road leading to the modern, quantum field theory interpretation of the Casimir effect. For the first time, it contains the local formulation of this effect in terms of the vacuum energy density and of the vacuum pressure. These authors derived the following explicit expression for the regularized energy-momentum (or stress-energy) tensor

$$\Theta^{\mu\nu} = -\frac{\pi^2}{180d^4}\left(\frac{1}{4}g^{\mu\nu} + z^\mu z^\nu\right), \qquad (5.15)$$

which was computed with the aid of an image-source construction of the corresponding Green's function. Here z^μ denotes a space-like four-vector orthogonal to the parallel plates, and d (as before) its separation.

Also for the first time, and although not explicitly stated, the calculations in [17] involved the zeta function procedure, in particular, Riemann and Epstein zeta functions. This method has evolved, starting from the subsequent seminal works of J.S. Dowker and R. Critchley (of 1976, Ref. [23] and of S. Hawking (of 1977, Ref. [24])—and incorporating a rather long list of contributions from different authors [10, 11, 14]—into a simple and mathematically elegant way of defining regularized vacuum energy densities in situations that nowadays very much generalize the original case considered by Casimir. Nowadays the zeta function regularization procedure pervades different aspects of quantum field theory (see in Chap. 1).

5.2.2 The Mystery of the Casimir Effect

Before introducing the concrete expressions which leads to the calculation of the Casimir effect from the point of view of the modern quantum field theory let us say a few words about an intriguing question that was posed before, namely, why is the Casimir effect less understood now than it was half a century ago? Why did it become a subject of more and more interest with the passage of the decades? Though the (tentative) answer to these questions will become more clear after the discussion below and once the actual calculations are performed, we can already point out in advance to the basic problem: unlike the van der Waals forces, which are always attractive, the ones appearing in the Casimir effect can be either attractive or repulsive. In the most simple case, the modern calculations about the Casimir effect indicate without doubt that if instead of two plates we considered the configuration formed by the two halves of a hollow sphere of a neutral perfectly conducting metallic material, if we would bring the two halves together in order to form a closed sphere, they would suffer a *repulsive* pressure. The sign of the Casimir force (and that of the vacuum energy density) is positive or negative depending crucially on the nature of the field (electromagnetic, i.e., the only one considered till now, scalar, etc.) and, for a given field, on the dimension of spacetime (usually four, but can be arbitrarily generalized), and, once the dimension fixed, on the particular geometry of the boundary and on the boundary conditions imposed on the field. Here, we are always talking about a flat spacetime but, of course, curved manifolds with different BC's come often into play.

One speaks nowadays of different (generalized) Casimir effects, as due to:

1. The nature of the background field in the vacuum.
2. The geometry of the boundary.
3. The boundary conditions.
4. The possible curvature of spacetime.

Summing up, already by closing in a trivial way the configuration considered by Casimir we obtain a repulsive pressure, which can in no way be explained as a kind of van der Waals-like force [146, 147]. On the other hand, for the multiple generalizations of the Casimir effect (to different fields, boundaries, dimensions, and spacetimes), the dependence of the sign of the force on them is anything but trivial. So is the *mystery* of the Casimir force born.

5.2.3 The Concept of the Vacuum Energy

It has been this generalization of the concept of the Casimir effect to incorporate all kind of contributions to the vacuum energy density in the situations described above what has rendered this concept so popular in quantum field theory . For an excellent review, much more detailed than the present *résumé* the lector is addressed to the report by Plunien, Müller and Greiner [130], which contains a detailed exposition of the developments in this field and, on its turn, 156 basic references. In particular, the Casimir effect can give rise to contributions to the surface tension of a curved conductor, can have cosmological consequences due to deviations from the Minkowskian geometry of spacetime, can lead to calculations of the self-energy for a scalar field confined to a cavity, or even to calculations in the bag model as a confining mechanism for quarks and gluons in QCD. It also gives the response of the vacuum to the presence of external fields. In this context Ambjørn and Wolfram [148] have discussed the vacuum energy of a charged scalar field in the presence of an external electric field. Finally, deviations form ideal conditions have also been the object of investigation, namely, non-zero temperature, non-infinitely conducting metals, plates of non-zero depth, etc. (for a sample of books, see [149–154].

Let us now describe the conceptual revolution brought about by the appearance of the Casimir force. As already remarked, at the beginning there was no difficulty in explaining the Casimir effect—and classical generalizations of it—by means of the Lifshitz's theory of the van der Waals forces. However, modern quantum field theory offers an alternative, general, more basic explanation from first principles of the Casimir force: it is due to a change of the vacuum energy, i.e., to a deviation of the zero-point energy caused by the presence of external constraints. In other words, Casimir's work stimulated investigations about the zero-point energy problem in quantum field theory which resulted in what is now commonly called the 'Casimir concept of the vacuum energy': the physical vacuum energy of a quantized field must be calculated with respect to its interaction with external constraints, and is thus defined as the difference between the zero-point energy corresponding to the vacuum configuration with constraints and to the free vacuum configuration, respectively. This formal definition must be supplemented, in general, with a regularization prescription in order to obtain a finite final expression. In this way a precise field quantization scheme starting from first principles is constructed (at least theoretically, in practice things are not so easy).

A first application of this concept of vacuum energy is an alternative calculation of the Casimir effect, which gives exactly the same result as the one obtained through its interpretation as due to retarded van der Waals forces. Actually, it was Casimir itself who showed for the first time that the zero-point energy of the electromagnetic field could successfully explain the van der Waals attraction. But the concept of vacuum (or Casimir) energy goes much further than his simple example and leads to results which in no way can be explained as due to van der Waals forces, such as the cases of repulsive pressure already mentioned. In the modern Casimir theory these forces arise and cannot be understood in the framework of conventional field theory, in which the zero-point energy is simply neglected. Two general methods for evaluation of the Casimir energy can be distinguished:

1. Summation of the series of energy eigenvalues corresponding to the zero-point field modes.
2. Determination of the vacuum stress-energy tensor in terms of local Green functions, obtained by constrained propagation of virtual field quanta.

Both methods should lead to the same result, but this is problematic because of the infinities involved and of the different regularization schemes used. Aside from this, specific (technical) difficulties appear in both cases (see the recent reference [155]). In the first, one is led (in principle) to calculate the whole energy spectrum for the free and for the constrained field modes, and this can only be achieved easily for simple geometries. In the second case, one has to determine the exact Green functions describing propagation in the presence of external boundaries. This is done by the usual image source construction, which is again easy only for simple geometries, or perturbatively by multiscattering expansions in the case of general constraints.

5.2.4 The Explicit, Regularized Definition of the Casimir Energy

For what has been said, it is now clear that a fundamental question connected with
the Casimir energy is the determination of the 'true' field Hamiltonian. In a model
without boundary conditions the Hamiltonian eigenvalue associated with the ground
state or vacuum (the zero-point energy) is always discarded because, in spite of being infinite, it can be reabsorbed in a suitable redefinition of the origin of energies.
The most popular way of putting such an adjustment into practice is normal ordering. Now, a most important implication of the concept of vacuum energy steaming
from the work of Casimir is the fact that the vacuum energy in quantum field theory
cannot be defined by means of normal ordering, since this procedure cannot possibly take into account the presence of arbitrary boundaries. The canonical formalism
of quantum field theory tells us that, for a scalar field of mass m, the Hamiltonian
operator takes the form

$$\hat{H} = \frac{1}{2} \sum_k \omega_k \left(a_k^\dagger a_k + a_k a_k^\dagger \right)$$

$$= \sum_k \omega_k \left(n_k - \frac{1}{2} \right)$$

$$= \sum_k \omega_k \left(a_k^\dagger a_k + \frac{1}{2} \right), \tag{5.16}$$

where $\omega_k^2 = k^2 + m^2$ are the eigenvalues of the Klein–Gordon operator, $a_k^\dagger$ and a_k
satisfy the canonical commutation relations for bosonic fields

$$\left[a_k, a_{k'}^\dagger \right] = \delta_{kk'}, \qquad [a_k, a_{k'}] = \left[a_k^\dagger, a_{k'}^\dagger \right] = 0, \tag{5.17}$$

and $n_k = a_k a_k^\dagger$ is the number operator, whose eigenvalues are non-negative integers.
   As the vacuum state $|0\rangle$ is defined by

$$a_k |0\rangle = 0, \tag{5.18}$$

when computing the vacuum expectation value

$$E_0 \equiv \langle 0 | \hat{H} | 0 \rangle \tag{5.19}$$

one gets a half of the sum over all the eigenfrequencies:

$$E_0 = \frac{1}{2} \sum_k \omega_k, \tag{5.20}$$

which is in general a divergent quantity. When instead of a bosonic field one considers a fermionic one the situation is similar. Only some signs are reversed, due to the presence of anticommutators in the place of the commutators above.[6]

5.2.5 Definition of the Casimir Energy Density and Its Relation with the Vacuum Energy

As we have just seen the Casimir energy is given by the expression (for the convenience of the reader, in this subsection we shall write everywhere the $\hbar$'s and c's explicitly)

$$E_{Cas} = \frac{\hbar}{2} \sum_n \omega_n. \tag{5.21}$$

In four-dimensional ultrastatic spacetime, the four-metric can be written as $g_4 = -(dx_0)^2 + g_3$. Correspondingly, we have $D_4 = -d_0^2 + D_3$ and $\omega_n = c\sqrt{\lambda_n(D_3)}$, λ_n being the eigenvalues of D_3. Then

$$E_{reg}(\epsilon) = \frac{1}{2}\hbar c \mu \sum_n \left(\frac{\lambda_n}{\mu^2}\right)^{\frac{1}{2}-\epsilon} = \frac{1}{2}\hbar c \mu \zeta_{D_3}\left(-\frac{1}{2} + \epsilon\right), \tag{5.22}$$

where

$$\zeta_{D_3}(s) = \frac{(4\pi)^{-d/2}}{\Gamma(s)} \left[\sum_{n=0}^{\infty} \frac{c_n}{s+n-d/2} + f(s) \right], \tag{5.23}$$

being $f(s)$ analytic and $c_n = \int_\Omega a_n + \int_{\partial\Omega} b_n$ (a volume and a surface integrals). $E_{reg}(\epsilon)$ is not analytic but meromorphic, with a pole (for $d=3$) at $\epsilon = 0$, of residue: $-\hbar c \mu c_2 (d=3)/(32\pi^2)$. The regulator can be suppressed only if $c_2 = 0$, i.e. for flat space with thin boundaries and for massless particles. If boundaries are plates (of surface S and separation L), then

$$E_{Cas}(L, S) = -\frac{\hbar c S}{L^3} \frac{\pi^2}{12} \zeta_R(-3), \tag{5.24}$$

where $\zeta_R = \zeta$ is the ordinary Riemann zeta function.

The pole can be absorbed into the bare action (term proportional to c_2), but there is no unique way to do this. The most economical one is through the principal part

---

[6]One should note that in the standard situations of QFT when no coupling of the vacuum diagrams is present, one can get rid of this term by a simple determination of the origin of energies using, e.g., the normal ordering prescription. An absolutely different case is when general relativity is considered in a quantum context, provided quantum vacuum fluctuations are taken to be a 'legal' form of energy (see later).

prescription (PP). This minimal subtraction yields

$$E_{Cas} \equiv \lim_{\epsilon \to 0} \frac{1}{2} \left[E_{reg}(+\epsilon) + E_{reg}(-\epsilon) \right] = \frac{1}{2} \hbar c \mu PP \zeta_{D_3}(-1/2), \qquad (5.25)$$

and is valid for simple poles [42]. To be mentioned is that: (i) μ can introduce a second (finite) ambiguity; (ii) it can be proven that the difference between the Casimir energy and the effective action to one loop is finite, independent of μ and proportional to the term c_2, of geometrical nature.

The one-loop effective action is given by

$$S_{eff} = \frac{1}{2} \ln \det D_4, \qquad (5.26)$$

where $D_4 = \partial_0^2 + D_3$ (Euclidean space). Then

$$\zeta_{D_4}(s) = \frac{\mu c T}{2\sqrt{\pi}} \frac{\Gamma(s - 1/2)}{\Gamma(s)} \zeta_{D_3} \left(s - \frac{1}{2} \right), \qquad (5.27)$$

with $T = \int dx/c$, the age of the universe. One obtains the following relation between E_{eff} and E_{Cas}

$$E_{eff} = E_{Cas} + \frac{1}{2} \hbar c \mu \left[\psi(1) - \psi \left(-\frac{1}{2} \right) \right] \frac{c_2}{(4\pi)^2}. \qquad (5.28)$$

Here μc_2 is independent of μ and $\psi(s) = \Gamma'(s)/\Gamma(s)$ is the digamma function.

Again, if $c_2 = 0$ then $E_{eff} = E_{Cas}$. On its turn, the vacuum energy is usually defined as

$$E_{vac} = \int \langle 0 | T_{00} | 0 \rangle, \qquad (5.29)$$

where $T_{\mu\nu}$ is the energy-momentum tensor, and is different (in general) both from E_{eff} and from E_{Cas}. If $T_{00} = T_{00}^{free}$, then $E_{vac} = E_{Cas}$. Yet another definition of E_{vac} is the following. Considering the total action (not just the action to one-loop), it is natural to define

$$E_{vac}^t = \frac{\Gamma_{eff}}{T}, \qquad (5.30)$$

what yields yet a different notion of vacuum energy. Summing up, we see that there are at least up to four different definitions of the energy corresponding to the vacuum, although all of them are closely related.

5.3 A Very Simple Computation of the Casimir Effect

As has been explained before, there are several (in the end equivalent) methods for obtaining the vacuum energy density corresponding to the Casimir effect. In this section, a detailed, pedagogical account will be given of the one which is based on the zeta-function regularization procedure. It will be shown how it yields (in many instances) exact results, without ever having to resort to the usual issue of infinity cancellations. This simple way of dealing—by analytic continuation—with the series of eigenmodes, can be made into a closed procedure owing to the theorem of zeta function regularization which deals with the commutation of the order of the summations of infinite series (see Chap. 2). The cases of a scalar field with Dirichlet, Neumann and periodic boundary conditions, and of an electromagnetic field between perfectly conducting plates—intimately related from a mathematical point of view—will be here investigated.

Next year we will celebrate the 65th anniversary of the publication by H.B.G. Casimir of the striking effect that now bears his name. As admitted by many physicists, it is (one of) the most beautiful and simple manifestation(s) of the striking vacuum structure of quantum field theory. It was shown first by Casimir—and checked experimentally later—that the vacuum energy density between two neutral, parallel, perfectly conducting plates of infinite extension is different from zero, when the vacuum energy in the absence of the plates is put equal to zero. In other words, the effect is simply due to the influence of the boundary conditions imposed on the vacuum configuration. With a different numerical value in each case, the effect also takes place when one imposes periodic, Dirichlet or Neumann boundary conditions on a Klein–Gordon field [148]. We will later specify the relations between these different situations.

In the case we have referred to (a single pair of parallel plates), the Casimir effect reveals itself in the form of an attractive force per unit area that tends to bring the plates together. However, it has been realized [130, 146–148] that by changing the boundary conditions, the Casimir force can also be made repulsive, the most simple cases being those of a sphere and of a closed box. However, these last are only theoretical results. They could not be checked experimentally till now, due in essence to the considerable difficulties involved in a material realization of this kind of boundary conditions.

Certainly, there are several (in the end equivalent) ways of obtaining the vacuum energy density corresponding to the Casimir effect [130]. The expression to be regularized is, as we know already, the (multiple) series of eigenvalues

$$\mathcal{E}_0 = \frac{1}{2} \sum_n \omega_n. \tag{5.31}$$

In principle, our procedure will be appropriate only in the case when all the n are known explicitly. But it is to be remarked that, as explained in the preceding chapters, even then the zeta-function regularization has been of limited use till recently, owing to the impossibility (in general) of performing a naive commutation of the order of the summations of the infinite series involved—and which do not happen to be absolutely convergent [102, 103]. It has been the correct calculation of the additional terms which appear when commuting these series—which was carried out in [48] for the first time—that began to transform the zeta-function regularization procedure into a powerful method which can deal directly with the infinite sums (5.31), as we shall here see.

In the first subsection the case of a massless scalar field in $\mathbb{S}^1 \times \mathbb{R}^d$ and $\mathbb{T}^2 \times \mathbb{R}^2$ spacetimes will be considered. The first is the most simple case while the second provides an example of a more complicated situation which also yields an exact result. Then the more general case of a massless scalar field with Dirichlet boundary conditions imposed on an arbitrary set of perpendicular pairs of parallel walls will be studied. Use is made of the results obtained before on the theorem of zeta-function regularization. Finally, in a last subsection we show how the results are immediately extensible to different physical situations, in particular to a massless scalar field with periodic or Neumann boundary conditions and to the original case of Casimir of an electromagnetic field in the interior of a set of perfectly conducting plates.

5.3.1 The Casimir Effect for a Free Massless Scalar Field in $\mathbb{S}^1 \times \mathbb{R}^d$ and in $\mathbb{T}^2 \times \mathbb{R}^2$ Spacetimes

1. *Spacetime* $\mathbb{S}^1 \times \mathbb{R}^d$, $d = 1, 2, 3, \ldots$

To start with, let us consider the case of a massless scalar field in $\mathbb{R}^{d+1}$ spacetime, $d = 1, 2, 3, \ldots$, satisfying periodic boundary conditions in only one of the space variables, i.e.

$$\varphi(x_1 + L, x_2, \ldots, x_d, t) = \varphi(x_1, x_2, \ldots, x_d, t). \tag{5.32}$$

The field satisfies the free Klein–Gordon equation

$$\Box \varphi(\vec{x}, t) = 0. \tag{5.33}$$

Let us call simply x the direction x_1, and $\vec{x}_T$ the rest (if any) of the spatial directions, $\vec{x}_T = (x_2, \ldots, x_d)$ [148]. The eigenmodes of the field are

$$\varphi(x, \vec{x}_T, t) = \exp\left(2\pi i n \frac{x}{L}\right) \exp(i\vec{k}_T \vec{x}_T) \exp(-i\omega_n t),$$

$$\omega_n = \sqrt{\vec{k}_T^2 + \left(\frac{2\pi n}{L}\right)^2}, \quad n \in \mathbb{Z}. \tag{5.34}$$

By directly adding together the contributions of all the eigenmodes, the total energy density (5.31) becomes

$$\mathcal{E}_0 = \frac{1}{(2\pi)^{d-1} 2L} \int_{\mathbb{R}^{d-1}} d\vec{k}_T \sum_{n=-\infty}^{\infty} \omega_n. \tag{5.35}$$

The zeta-function regularization procedure begins with the replacement of the exponent $1/2$ in ω_n with $-s/2$:

$$\mathcal{E}_0(s) = \frac{1}{(2\pi)^{d-1}2L} \frac{2\pi^{(d-1)/2}}{\Gamma((d-1)/2)} \int_0^\infty dk\, k^{d-2} \sum_{n=-\infty}^\infty \left[k^2 + \left(\frac{2\pi n}{L}\right)^2\right]^{-s/2},$$

(5.36)

where s is such that $\mathrm{Re}(s) > 0$ is big enough in order that the k-integration provides a finite result

$$\mathcal{E}_0(s) = \frac{2^{-s}\pi^{(d-1)/2-s}}{L^{d-s-2}} \frac{\Gamma((s-d+1)/2)}{\Gamma(s/2)} \sum_{n=1}^\infty n^{d-s+1}.$$

(5.37)

The analytic continuation from these values of s to the one, $s = -1$, we are interested in, is provided by the Riemann zeta function (Chap. 1). Using then the reflection formula (analytical continuation), we obtain the following closed expression for the zeta-function-regularized vacuum energy density

$$\mathcal{E} = -\frac{\Gamma((d+1)/2)\zeta(d+1)}{(\sqrt{\pi}L)^{d+1}}.$$

(5.38)

In particular, for $d = 1, 2, 3$ we obtain, respectively,

$$\mathcal{E}(d=1) = -\frac{\pi}{6L^2}, \qquad \mathcal{E}(d=2) = -\frac{\zeta(3)}{2\pi L^3}, \qquad \mathcal{E}(d=3) = -\frac{\pi^2}{90L^4}. \quad (5.39)$$

2. *Spacetime* $\mathbb{T}^2 \times \mathbb{R}^2$

In this case the field satisfies periodic boundary conditions in two of the spatial variables

$$\varphi(x_1 + L, x_2 + L_2, x_3, t) = \varphi(x_1, x_2, x_3, t).$$

(5.40)

The formal vacuum energy density is now

$$\mathcal{E}_0 = \frac{1}{4\pi L_1 L_2} \int_{-\infty}^\infty dk \sum_{n_1,n_2=-\infty}^\infty \left[k^2 + \left(\frac{2\pi n_1}{L_1}\right)^2 + \left(\frac{2\pi n_2}{L_2}\right)^2\right]^{1/2}, \quad (5.41)$$

where L_1 and L_2 are the lengths of the circumferences of the torus $\mathbb{T}^2$. As before, let us consider

$$\mathcal{E}_0(s) = \frac{1}{2\pi L_1 L_2} \int_0^\infty dk \sum_{n_1,n_2=-\infty}^\infty \left[k^2 + \left(\frac{2\pi n_1}{L_1}\right)^2 + \left(\frac{2\pi n_2}{L_2}\right)^2\right]^{-s/2}, \quad (5.42)$$

for s such that $\mathrm{Re}(s) > 0$ is big enough so that the k-integration be convergent:

$$\mathcal{E}_0(s) = \frac{2^{-s-1}\pi^{1/2-s}}{L_1 L_2} \frac{\Gamma((s-1)/2)}{\Gamma(s/2)} \sum_{n_1,n_2=-\infty}^\infty \left[\left(\frac{n_1}{L_1}\right)^2 + \left(\frac{n_2}{L_2}\right)^2\right]^{(1-s)/2}. \quad (5.43)$$

The analytic continuation of this double series to $s = -1$ is readily done (see point 2 of the following section). Using Jacobi's identity twice (see Chap. 2), one obtains [48, 51] one obtains

$$E_2(s; a_1, a_2) = -\frac{1}{2}\left(a_1^{-s} + a_2^{-s}\right)\zeta(2s)$$

$$+ \frac{\sqrt{\pi}}{2}\frac{a_1^{1-s} + a_2^{1-s}}{\sqrt{a_1 a_2}}\frac{\Gamma(s - 1/2)}{\Gamma(s)}\zeta(2s - 1)$$

$$+ \frac{\pi^{2s-1}}{\sqrt{a_1 a_2}}\frac{\Gamma(1 - s)}{\Gamma(s)}E_2\left(1 - s; \frac{1}{a_1}, \frac{1}{a_2}\right), \quad a_1, a_2 > 0, \quad (5.44)$$

or, explicitly, in terms of the modified Bessel function K_ν

$$E_2(s; a_1, a_2) = -\frac{1}{2}a_2^{-s}\zeta(2s) + \frac{1}{2}a_2^{1/2-s}\left(\frac{\pi}{a_1}\right)^{1/2}\frac{\Gamma(s - 1/2)}{\Gamma(s)}\zeta(2s - 1)$$

$$+ \frac{2\pi^2}{\Gamma(s)}a_1^{-s/2-1/4}a_2^{-s/2+1/4}$$

$$\cdot \sum_{n_1, n_2=1}^{\infty} n_1^{s-1/2}n_2^{-s+1/2}K_{s-1/2}(2\pi\sqrt{a_2/a_1}n_1 n_2). \quad (5.45)$$

In order to simplify things, let us restrict ourselves to the particular case $L_1 = L_2 \equiv L$. We get [17, 105–107]

$$\sum_{n_1, n_2=-\infty}^{\infty} \left(n_1^2 + n_2^2\right)^{(1-s)/2} = 4\zeta\left(\frac{s - 1}{2}\right)\beta\left(\frac{s - 1}{2}\right), \quad (5.46)$$

where the beta function $\beta(s)$ is defined by [100]

$$\beta(s) = \sum_{n=0}^{\infty}(-1)^n(2n + 1)^{-s}. \quad (5.47)$$

Now, making use of the reflection formula for this beta function

$$\beta(s) = \left(\frac{\pi}{2}\right)^{s-1}\cos\left(\frac{\pi s}{2}\right)\Gamma(1 - s)\beta(1 - s), \quad (5.48)$$

for the regularized Casimir energy density, we obtain in the present case

$$\mathcal{E} = -\frac{\beta(2)}{3L^4} = -0.30532186L^{-4}. \quad (5.49)$$

Notice that this is, again, a closed result, in terms of a particular value of the beta function.

Before proceeding with the generalization to $\mathbb{T}^2 \times \mathbb{R}^{d-1}$, $d = 2, 3, 4, \ldots$, we shall consider first the case of Dirichlet boundary conditions. The reason is that, being then the spectrum of eigenvalues n limited to $n = 1, 2, 3, \ldots$, this is a more immediate situation from the point of view of zeta-function regularization. The formulas corresponding to the periodic case will then be obtained through simple combinatorics.

5.3.2 The Case of a Massless Scalar Field Between p Perpendicular Pairs of Parallel Walls with Dirichlet Boundary Conditions

The walls are, in mathematical terms, $(d-1)$-dimensional hyperplanes, of equations

$$x_1 = 0, \quad x_1 = L_1; \qquad x_2 = 0, \quad x_2 = L_2; \quad \ldots; \quad x_p = 0, \quad x_p = L_p. \tag{5.50}$$

The Dirichlet boundary conditions consist in the annihilation of the field at each of these boundaries W_j, $j = 1, \ldots, p$

$$\varphi(\vec{x}, t) = 0, \quad \vec{x} \in W_j, \ j = 1, \ldots, p. \tag{5.51}$$

The field modes are now

$$\varphi(x_1, \ldots, x_p, \vec{x}_T, t) = \prod_{j=1}^{p} \sin\left(\frac{\pi n_j x_j}{L_j}\right) \exp(i\vec{k}_T \vec{x}_T) \exp(-i\omega_n t),$$

$$\omega_n = \sqrt{\vec{k}_T^2 + \sum_{j=1}^{p} \left(\frac{\pi n_j}{L_j}\right)^2}, \quad n_j = 1, 2, 3, \ldots; j = 1, \ldots, p,$$

$$\tag{5.52}$$

and the (formal) unregularized vacuum energy density is given by

$$\mathcal{E}_0^D = \frac{(2\pi)^{p-d}}{2 \prod_{j=1}^{p} L_j} \int_{\mathbb{R}^{d-p}} d\vec{k}_T \sum_{n_1, \ldots, n_p = 1}^{\infty} \left[\vec{k}_T^2 + \sum_{j=1}^{p} \left(\frac{\pi n_j}{L_j}\right)^2\right]^{1/2}. \tag{5.53}$$

Define as before ((5.36) and (5.42))

$$\mathcal{E}_0^D(s) = \frac{(2\pi)^{p-d}}{2 \prod_{j=1}^{p} L_j} \frac{2\pi^{(d-p)/2}}{\Gamma((d-p)/2)} \int_0^{\infty} dk\, k^{d-p-1}$$

$$\cdot \sum_{n_1, \ldots, n_p = 1}^{\infty} \left[k^2 + \sum_{j=1}^{p} \left(\frac{\pi n_j}{L_j}\right)^2\right]^{-s/2}, \tag{5.54}$$

for s with $\text{Re}(s) > 0$ big enough in order to obtain a convergent integral in k. After calculating it, we get

$$\mathcal{E}_0^D(s) = \frac{2^{p-d-1}\pi^{(d-p)/2-s}}{\prod_{j=1}^p L_j} \frac{\Gamma((s-d+p)/2)}{\Gamma(s/2)} \sum_{n_1,\ldots,n_p=1}^{\infty} \left[\sum_{j=1}^p \left(\frac{n_j}{L_j}\right)^2\right]^{(d-s-p)/2}.$$

(5.55)

1. Particular Case $p = d - 1$

Let us first consider the particular case $p = d - 1$ (of which case 2 of the preceding subsection is an example). We obtain

$$\mathcal{E}_0^D(s) = \frac{\pi^{1/2-s}}{4\prod_{j=1}^p L_j} \frac{\Gamma((s-1)/2)}{\Gamma(s/2)} \sum_{n_1,\ldots,n_p=1}^{\infty} \left[\sum_{j=1}^p \left(\frac{n_j}{L_j}\right)^2\right]^{(1-s)/2}.$$

(5.56)

Corresponding to that in Sect. 5.2, let us first consider the case $d = 3$, $p = 2$. We shall relate the double sum here with the one in (5.43) and (5.46), namely (for $L_1 = L_2 \equiv L$)

$$\Sigma_1\left(\frac{1-s}{2}\right) \equiv \sum_{n_1,n_2=-\infty}^{\infty} (n_1^2 + n_2^2)^{(1-s)/2} = 4\zeta(s-1) + 4\sum_{n_1,n_2=1}^{\infty} (n_1^2 + n_2^2)^{(1-s)/2}.$$

(5.57)

Calling the last series $\Sigma_2(s)$, for its analytic continuation to $s = -1$ we obtain

$$\Sigma_2(-1) = \sum_{n_1,n_2=1}^{\infty} (n_1^2 + n_2^2) = \frac{1}{2}\Sigma_1(-1) - \zeta(-2) = \zeta(-1)\beta(-1).$$

(5.58)

Finally, use of the reflection formulas yields for this case:

$$\mathcal{E}^D(d=3, p=2) = \frac{1}{16L^4}\left[\frac{\zeta(3)}{\pi} - \frac{\beta(2)}{3}\right] = 0.00483155L^{-4}.$$

(5.59)

At this point, in order to express (as in (5.57), (5.58)) the multiple sum (5.56) in terms of the zeta function, for arbitrary dimension d, we shall make use of the results on the commutation of the order of the summations of infinite, non-absolutely convergent series (the zeta-function regularization theorem of Chap. 2). They are equally valid in the general case: $d \geq p$ arbitrary, so let us turn to the next point where we consider the general situation.

2. Zeta-Function Regularization of Multiple Series

Those we are interested in here are always of the form

$$E_p(s; a_1, \ldots, a_p) \equiv \sum_{n_1,\ldots,n_p=1}^{\infty} (a_1 n_1^2 + \cdots + a_p n_p^2)^{-s}.$$

(5.60)

We now recall the zeta-function regularization theorem, which, in the quadratic case, just renders the Jacobi identity for the theta function θ_3 (for many more details on this point see Chap. 2). As a consequence of the theorem, in the general case, we get the following recurrence (that extends the previous on valid for $p = 2$):

$$
\begin{aligned}
E_p(s; a_1, \ldots, a_p) = {}& -\frac{1}{2} E_{p-1}(s; a_2, \ldots, a_p) \\
& + \frac{1}{2} \sqrt{\frac{\pi}{a_1}} \frac{\Gamma(s - 1/2)}{\Gamma(s)} E_{p-1}(s - 1/2; a_2, \ldots, a_p) \\
& + \frac{\pi^s}{\Gamma(s)} a_1^{-s/2} \sum_{k=0}^{\infty} \frac{a_1^{k/2}}{k!(16\pi)^k} \prod_{j=1}^{k} [(2s - 1)^2 - (2j - 1)^2] \\
& \cdot \sum_{n_1, \ldots, n_p = 1}^{\infty} n_1^{s-k-1} \left(a_2 n_2^2 + \cdots + a_p n_p^2 \right)^{-(s+k)/2} \\
& \cdot \exp\left[-\frac{2\pi}{\sqrt{a_1}} n_1 \left(a_2 n_2^2 + \cdots + a_p n_p^2 \right)^{1/2} \right], \quad a_1, \ldots, a_p > 0.
\end{aligned}
$$
(5.61)

These expressions, (5.45) and (5.61), provide the analytic continuation of (5.60) to all values of s. We must not be confused by the imposing aspect of the last term in (5.45) and (5.61). This term converges very quickly and amounts only to a small correction to be added to the first two terms, in each case. In a numerical calculation up to, say, 6 or 8 decimal places, only the first couple of summands of the last series need to be taken into account (see Chap. 4 for precise details). Thus, in practice, (5.61) can be viewed as a recurrent equation with a correction piece Δ (the last term) which is very small and can be estimated numerically with good approximation.

3. General Case with Dirichlet Boundary Conditions

Bringing together the results obtained here, it is now not difficult to write the general formulas for the Casimir energy density of a scalar field in $\mathbb{R}^{d+1}$ spacetime, with Dirichlet boundary conditions imposed on p perpendicular pairs of parallel walls. We need only to introduce (5.61) or (5.45) into (5.55) and set $s = -1$, that is

$$
\mathcal{E}^D(d, p) = -\frac{\pi^{(d-p+1)/2}}{2^{d-p+2}} \frac{\Gamma((p - d - 1)/2)}{\prod_{j=1}^{p} L_j} E_p\left((p - d - 1)/2; L_1^{-2}, \ldots, L_p^{-2}\right).
$$
(5.62)

with E_p given in (5.61), (5.45).

In the particular case of point 1, $p = d - 1$, the k-series in (5.61) actually reduces to the first two terms only, i.e. $k = 0$ and $k = 1$. The recurrence (5.61) simplifies to

$$
E_p\left(-1; L_1^{-2}, \ldots, L_p^{-2}\right)
$$

$$= -\frac{1}{2} E_{p-1}\left(-1; L_2^{-2}, \ldots, L_p^{-2}\right)$$

$$+ \frac{2}{3} \frac{\pi L_1}{\Gamma(-1)} E_{p-1}\left(-3/2; L_2^{-2}, \ldots, L_p^{-2}\right)$$

$$+ \frac{1}{\pi L_1 \Gamma(-1)} \sum_{n_1,\ldots,n_p=1}^{\infty} n_1^{-2} \left\{ \frac{1}{2\pi L_1 n_1} + \left[\left(\frac{n_2}{L_2}\right)^2 + \cdots + \left(\frac{n_p}{L_p}\right)^2\right]^{1/2} \right\}$$

$$\cdot \exp\left\{ -2\pi L_1 n_1 \left[\left(\frac{n_2}{L_2}\right)^2 + \cdots + \left(\frac{n_p}{L_p}\right)^2\right]^{1/2} \right\},$$

$$L_1 \geq L_2 \geq \cdots \geq L_p. \tag{5.63}$$

Notice that $E_p(-1; a_1, \ldots, a_p) = 0$, $a_1, \ldots, a_p > 0$. This is clear from (5.63), due to the presence of the $\Gamma(-1)$ factor in the denominators. However, in (5.62), for $p = d - 1$ this factor just compensates the $\Gamma((p - d - 1)/2)$ one in the numerator, thus providing a finite (regularized) result. The same is true in general for the cases when $p = d - (2h + 1)$, $h = 0, 1, 2, \ldots$. One has

$$E_p(-h - 1; a_1, \ldots, a_p) = 0, \quad a_1, \ldots, a_p > 0; \ h = 0, 1, 2, \ldots, \tag{5.64}$$

owing to a factor $1/\Gamma(-h - 1)$ which compensates in (5.62), being the final result finite and non-zero.

On the other hand, for $p = d - 2h$, $h = 0, 1, 2, \ldots$, this situation only takes place at the second term of the r.h.s. of (5.62), where the factor $\Gamma(-h - 1)$ compensates a similar factor at the denominator of $E_{p-1}(-h - 1)$.

As for an explicit calculation, let us consider the example $d = 4$, $p = 3$, next to the one studied at point 1 and which cannot be dealt with directly with the help of Epstein zeta functions [17, 105–107] (because no formula for $Z_3(s)$ in terms of simple Dirichlet series exists). In order to make the derivation easier, let us put $L_1 = L_2 = L_3 \equiv L$. Then (5.63) reduces to

$$E_3(-1) = -\frac{1}{2} E_2(-1) + \frac{2}{3} \frac{\pi}{\Gamma(-1)} E_2(-3/2)$$

$$+ \frac{1}{\pi \Gamma(-1)} \sum_{n_1, n_2, n_3=1}^{\infty} n_1^{-2} \left(\frac{1}{2\pi n_1} + \sqrt{n_2^2 + n_3^2}\right)$$

$$\cdot \exp\left(-2\pi n_1 \sqrt{n_2^2 + n_3^2}\right), \tag{5.65}$$

where a common factor L^{-2} has been taken off. This $E_2(s)$ just coincides now with the $\Sigma_2(s)$ introduced between (5.57) and (5.58)

$$E_2(s) = \Sigma_2(s) = \zeta(s)\beta(s) - \zeta(2s). \tag{5.66}$$

Substituting into (5.62), we obtain

$$\mathcal{E}^D(4,3) = -\frac{1}{32L^5}\left[\frac{\zeta(3)}{\pi} - \frac{\beta(2)}{3} + \frac{3}{2\pi^2}\zeta(5/2)\beta(5/2) - \frac{\pi^2}{45} + 4S\right], \quad (5.67)$$

where S is the series in (5.65). It is immediate to check that the contribution of the S-term in (5.67) is of order 10^{-5}, the full numerical result being

$$\mathcal{E}^D(4,3) = -0.0023L^{-5}. \quad (5.68)$$

5.3.3 Massless Scalar Field with Periodic and Neumann Boundary Conditions, and Electromagnetic Field

As we shall see, all these cases are obtained from the Dirichlet one considered in Sect. 5.2.

1. *Scalar Field with Neumann Boundary Conditions*

The spacetime will be $\mathbb{R}^{d+1}$ and the Neumann boundary conditions are imposed on a set of p perpendicular pairs of parallel walls W_j, $j = 1, \ldots, p$ (the same as in Sect. 5.2, (5.50))

$$\partial_{\vec{x}}\varphi(\vec{x},t) = 0, \quad \vec{x} \in W, \ j = 1, \ldots, p. \quad (5.69)$$

The eigenmodes are

$$\varphi(x_1, \ldots, x_p, \vec{x}_T, t) = \prod_{j=1}^{p} \sin\left(\frac{\pi n_j x_j}{L_j}\right) \exp(i\vec{k}_T\vec{x}_T) \exp(-i\omega_n t),$$

$$\omega_n = \sqrt{\vec{k}_T^2 + \sum_{j=1}^{p}\left(\frac{\pi n_j}{L_j}\right)^2}, \quad n_j = 1, 2, 3, \ldots, \ j = 1, \ldots, p,$$

$$(5.70)$$

and the vacuum energy density

$$\mathcal{E}_0^N = \frac{(2\pi)^{p-d}}{2\prod_{j=1}^{p}L_j}\int_{\mathbb{R}^{d-p}} d\vec{k}_T \sum_{n_1,\ldots,n_p=0}^{\infty}\left[\vec{k}_T^2 + \sum_{j=1}^{p}\left(\frac{\pi n_j}{L_j}\right)^2\right]^{1/2}, \quad (5.71)$$

which leads to (cf. (5.55))

$$\mathcal{E}_0^N(s) = \frac{2^{p-d-1}\pi^{(d-p)/2-s}}{\prod_{j=1}^{p}L_j}\frac{\Gamma((s-d+p)/2)}{\Gamma(s/2)}\sum_{n_1,\ldots,n_p=0}^{\infty}{}'\left[\sum_{j=1}^{p}\left(\frac{n_j}{L_j}\right)^2\right]^{(d-s-p)/2}, \quad (5.72)$$

where the prime means that the term $n_1 = \cdots = n_p = 0$ is to be excluded from the summation. As we see, $\mathcal{E}_0^N$ differs from $\mathcal{E}_0^D$, (5.53), only in the range of the series on the n_j's, that now extend from $0, 1, 2, \ldots$ to infinity (instead of from $1, 2, \ldots$ to infinity as in the Dirichlet case), and the same with (5.72) and (5.55). The analytic continuation of $\mathcal{E}_0^N(s)$ to $s = -1$, which provides the regularized vacuum energy density $\mathcal{E}^N(d, p)$ in the Neumann case, is correspondingly obtained from the Dirichlet result $\mathcal{E}^D(d, p)$ by simple combinatorics involving also the values of $\mathcal{E}^D(d', p')$ for $d' < d, p' < p$.

In fact, let us consider the identity

$$
\sideset{}{'}\sum_{n_1,\ldots,n_p=0}^{\infty} \left(a_1 n_1^2 + \cdots + a_p n_p^2\right)^r
$$

$$
= \sum_{k=0}^{p-1} \sum_{C(p,k)} \sum_{n_1,\ldots,n_{j_1-1},n_{j_1+1},\ldots,n_{j_k-1},n_{j_k+1},\ldots,n_p=1}^{\infty} \left(a_1 n_1^2 + \cdots + a_{j_1-1} n_{j_1-1}^2\right.
$$

$$
\left. + a_{j_1+1} n_{j_1+1}^2 + \cdots + a_{j_k-1} n_{j_k-1}^2 + a_{j_k+1} n_{j_k+1}^2 + \cdots + a_p n_p^2\right)^r, \quad (5.73)
$$

where the sum over $C(p, k)$ means sum over the $\binom{p}{k}$ selections of k indices $1 \leq j_1 < \cdots < j_k \leq p$ among the indices $1, \ldots, p$. It is now immediate, using (5.73), that

$$
\mathcal{E}_{L_1,\ldots,L_p}^N(d, p)
$$

$$
= \sum_{k=0}^{p-1} \sum_{C(p,k)} \frac{1}{\prod_{l=1}^{k} L_{j_l}} \mathcal{E}_{L_1 \ldots L_{j_1-1} L_{j_1+1} \ldots L_{j_k-1} L_{j_k+1} \ldots L_p}^D (d-k, p-k).
$$

$$
(5.74)
$$

If all the L_j, $j = 1, \ldots, p$, are equal, this formula simplifies to

$$
\mathcal{E}^N(d, p) = \sum_{k=0}^{p-1} \binom{p}{k} L^{-k} \mathcal{E}^D(d-k, p-k). \tag{5.75}
$$

2. Scalar Field with Periodic Boundary Conditions

As before we consider, for simplicity, the massless case. With a little more effort the corresponding massive case can be treated by exactly the same procedure. Let now $L_1, \ldots, L_p$ denote the lengths of the circumferences of the hypertorus $\mathbb{T}^p$. The spacetime is $\mathbb{T}^p \times \mathbb{R}^{d-p+1}$, i.e. the field satisfies periodic boundary conditions on p coordinates

$$
\varphi(x_1 + L, \ldots, x_p + L_p, x_{p+1}, \ldots, x_d, t) = \varphi(x_1, \ldots, x_d, t). \tag{5.76}
$$

The eigenmodes of the field are given by

$$
\varphi(x_1, \ldots, x_p, \vec{x}_T, t) = \prod_{j=1}^{p} \exp\left(2\pi i \frac{n_j x_j}{L_j}\right) \exp(i\vec{k}_T \vec{x}_T) \exp(-i\omega_n t),
$$

$$
\omega_n = \sqrt{\vec{k}_T^2 + \sum_{j=1}^{p} \left(\frac{2\pi n_j}{L_j}\right)^2}, \quad n_j \in \mathbb{Z}; \; j = 1, \ldots, p, \tag{5.77}
$$

and the vacuum energy density is

$$
\mathcal{E}_0^P = \frac{(2\pi)^{p-d}}{2\prod_{j=1}^{p} L_j} \int_{\mathbb{R}^{d-p}} d\vec{k}_T \sum_{n_1,\ldots,n_p=-\infty}^{\infty} \left[\vec{k}_T^2 + \sum_{j=1}^{p}\left(\frac{2\pi n_j}{L_j}\right)^2\right]^{1/2}, \tag{5.78}
$$

the difference with (5.53) being now—apart from the range of the summation indices n_j of the series—the fact that the L_j's accompanying these indices are here $L_j/2$ (as compared with those of (5.53)). We get

$$
\mathcal{E}_0^P(s) = \frac{2^{-s-1}\pi^{(d-p)/2-s}}{\prod_{j=1}^{p} L_j} \frac{\Gamma((s-d+p)/2)}{\Gamma(s/2)} \sum_{n_1,\ldots,n_p=1}^{\infty} \left[\sum_{j=1}^{p}\left(\frac{n_j}{L_j}\right)^2\right]^{(d-s-p)/2}. \tag{5.79}
$$

Notice an additional factor 2^{d-s-p} with respect to the result (5.55). Moreover, from

$$
\sideset{}{'}\sum_{n_1,\ldots,n_p=-\infty}^{\infty} \left(a_1 n_1^2 + \cdots + a_p n_p^2\right)^r
$$

$$
= \sum_{k=0}^{p-1} \sum_{C(p,k)} 2^{p-k} \sum_{n_1,\ldots,n_{j_1-1},n_{j_1+1},\ldots,n_{j_k-1},n_{j_k+1},\ldots,n_p=1}^{\infty} \left(a_1 n_1^2 + \cdots + a_{j_1-1} n_{j_1-1}^2\right.
$$

$$
\left. + a_{j_1+1} n_{j_1+1}^2 + \cdots + a_{j_k-1} n_{j_k-1}^2 + a_{j_k+1} n_{j_k+1}^2 + \cdots + a_p n_p^2\right)^r, \tag{5.80}
$$

with the same meaning as in (5.73), we obtain

$$
\mathcal{E}_{L_1,\ldots,L_p}^P(d, p) = \sum_{k=0}^{p-1} \sum_{C(p,k)} \frac{2^{d-k+1}}{\prod_{l=1}^{k} L_{j_l}} \mathcal{E}_{L_1\ldots L_{j_1-1} L_{j_1+1}\ldots L_{j_k-1} L_{j_k+1}\ldots L_p}^D(d-k, p-k). \tag{5.81}
$$

In the particular case $L_1 = \cdots = L_p \equiv L$, this reduces to

$$
\mathcal{E}^P(d, p) = \sum_{k=0}^{p-1} \binom{p}{k} 2^{d-k+1} L^{-k} \mathcal{E}^D(d-k, p-k). \tag{5.82}
$$

3. *Electromagnetic Field Between p Perpendicular Pairs of Neutral, Perfectly Conducting Plates*

This case can also be related with the above ones, in particular with the Dirichlet case for a massless scalar field. Actually, the relations that we have obtained before, (5.74) and (5.81), and the one we shall derive in this point, are reversible (see [148] for the inverse relations) and allow us to connect in both directions any two of the cases that we have studied.

For a massless vector field in $\mathbb{R}^{d+1}$ spacetime in the presence of p perpendicular pairs of parallel plates—neutral, infinite and perfectly conducting—given mathematically by the same equations as before, (5.50), the boundary conditions are

$$n^{\mu} F^{*}_{\mu v_1 \dots v_{d-2}} = 0, \tag{5.83}$$

where F^* is the dual of the field-strength tensor: $F^{*}_{\mu_1 \dots \mu_{d-1}} = \epsilon_{\mu_1 \dots \mu_{d-1} v \lambda} F^{v \lambda}$. By introducing the potential A_μ such that $F_{\mu v} = \partial_\mu A_v - \partial_v A_\mu$, and working in the radiation gauge

$$A_0 = 0, \qquad \partial_1 A_1 + \dots + \partial_d A_d = 0, \tag{5.84}$$

the eigenvalues of the field are found to be given by [148]

$$A_j = L_j \cos(k_j x_j) \prod_{l=1, l \neq j}^{p} \sin(k_l x_l) \exp\left[i(\vec{k}_T \vec{x}_T - \omega t)\right], \quad j = 1, \dots, p,$$

$$A_h = ib_h \prod_{l=1}^{p} \sin(k_l x_l) \exp\left[i(\vec{k}_T \vec{x}_T - \omega t)\right], \quad h = p+1, \dots, d, \tag{5.85}$$

$$k_j = \frac{\pi n_j}{L_j}, \quad \vec{k}_T^2 = \omega^2 - \sum_{j=1}^{p} \vec{k}_j^2, \quad n_j = 0, 1, 2, \dots.$$

The gauge condition (5.84) implies

$$L_j k_j + \vec{b} \cdot \vec{k}_T = 0 \tag{5.86}$$

and forbids modes for which two or more of the n_j vanish (because this would imply $A_j = 0$, $j = 1, \dots, p$). The vacuum energy density is the same as in the Dirichlet case, (5.53), but for the range of the multiple series, which now extend from zero to infinity with the restriction (5.86).

For the corresponding function $\mathcal{E}^{em}(s)$, we get

$$\mathcal{E}^{em}(s) = \frac{2^{p-d-1} \pi^{(d-p)/2-s}}{\prod_{j=1}^{p} L_j} \frac{\Gamma((s-d+p)/2)}{\Gamma(s/2)} \sideset{}{''}\sum_{n_1, \dots, n_p = 0}^{\infty} \left[\sum_{j=1}^{p} \left(\frac{n_j}{L_j}\right)^2\right]^{(d-s-p)/2}.$$

$$\tag{5.87}$$

The double prime means "restricted by (5.86)", that is, there are $d - 1$ independent components $A_j \neq 0$ if all the $n_j \neq 0$, and there remains only one component $A_j \neq 0$ if one of the $n_j = 0$.

This can now be related with the result for a massless scalar field with Dirichlet boundary conditions, (5.55). Making use of an identity between multiseries for the present case similar to (5.73) and analytically continuing the expression to $s = -1$, we obtain the formula

$$\mathcal{E}^{em}_{L_1 \ldots L_p}(d, p) = (d - 1)\mathcal{E}^D_{L_1 \ldots L_p}(d, p) + \sum_{j=1}^{p} \frac{1}{L_j}\mathcal{E}^D_{L_1 \ldots L_{j-1}L_{j+1} \ldots L_p}(d - 1, p - 1).$$

(5.88)

This is the relation we were looking for between the scalar Dirichlet and the electromagnetic vacuum energy densities.

As for a particular example of the complete set of connections that one can derive among all the different boundary conditions for fields of different spin, we shall now obtain the relation existing between the electromagnetic case and the periodic one for a scalar field. To be concrete, for $d = 3$ and $p = 2$, using (5.81) and (5.88), we have

$$\mathcal{E}^P_{L_1 L_2}(3, 2) = 16\mathcal{E}^D_{L_1 L_2}(3, 2) + 8\left[\frac{1}{L_1}\mathcal{E}^D_{L_2}(2, 1) + \frac{1}{L_2}\mathcal{E}^D_{L_1}(2, 1)\right],$$

$$\mathcal{E}^{em}_{L_1 L_2}(3, 2) = 2\mathcal{E}^D_{L_1 L_2}(3, 2) + \frac{1}{L_1}\mathcal{E}^D_{L_2}(2, 1) + \frac{1}{L_2}\mathcal{E}^D_{L_1}(2, 1) \qquad (5.89)$$

$$= \frac{1}{8}\mathcal{E}^P_{L_1 L_2}(3, 2).$$

Numerically, from (5.59) we get, for $L_1 = L_2 \equiv L$,

$$\mathcal{E}^D(2, 1) = -\frac{\zeta(3)}{16\pi L^3} = -0.02391416L^{-3},$$

$$\mathcal{E}^P(3, 2) = 16\left[\mathcal{E}^D(3, 2) + \frac{1}{L}\mathcal{E}^D(2, 1)\right] = -0.30532176L^{-4}, \qquad (5.90)$$

$$\mathcal{E}^{em}(3, 2) = -0.03816522L^{-4}.$$

The last one is the value of the Casimir energy density corresponding to the electromagnetic field between two perpendicular pairs of neutral, infinite, perfectly conducting, parallel plates in $3 + 1$ spacetime dimensions. Again it is remarkable that all these results are exactly given in terms of values of well-known special functions at some integer points (5.89). In this way, the 8-decimal precision adopted in (5.90) can actually be made arbitrarily large.

Chapter 6
Five Physical Applications of the Inhomogeneous Generalized Epstein–Hurwitz Zeta Functions

In this chapter some explicit applications to the regularization, by means of Hurwitz zeta-functions considered in previous chapters, of different problems which have appeared recently in the physical literature, are considered. This kind of zeta functions show up profusely in different applications of quantum physics where regularization techniques are needed, in particular, when one deals with a massive quantum field theory in a (totally or partially) compactified spacetime (spherical or toroidal compactification, for instance). Aside from the interest that a detailed mathematical study of these functions may have on its own (e.g. in number theory), what is actually needed for most physical applications—as we shall explicitly see later—is always the numerical value of these functions, and of their derivatives with respect to the variable s and to the parameter a, for negative (half)integer values of s and for a few simple fractional values of a (like $a = 1/2, 1/4, 3/4, \ldots$). Concerning the parameter b^2, usually we need the large- or small-b^2 behavior of the series only.

The final results involve analytical continuation of the functions and also a principal part prescription in order to deal with (possible) poles. This procedure has already been checked in several situations, see [42, 126]. The results will be always given in terms of sums of Hurwitz zeta functions and generally under the form of asymptotic expansions.

Actually, this situation has been described in detail in Chap. 4 already. In the present one several, quite different applications will be presented (five of them, in all). A first direct application is described in the first section to the calculation of the Casimir energy corresponding to compact universes without boundary. In the following one a second application is considered, namely the calculation of the sum over one-loop integrals which yields the cross section of a scattering process in a Kaluza–Klein model with spherical compactification. Another application is the study of the critical behavior of a field theory at non-zero temperature. As the fourth example of this chapter the quantization of two-dimensional gravity by means of the Wheeler–De Witt equation is discussed. Finally, the last case considered is the use of the spectral zeta function for both scalar and vector fields on a spacetime with a noncommutative toroidal part.

E. Elizalde, *Ten Physical Applications of Spectral Zeta Functions*,
Lecture Notes in Physics 855,
DOI 10.1007/978-3-642-29405-1_6, © Springer-Verlag Berlin Heidelberg 2012

6.1 Application: The Casimir Energy over Riemann Surfaces

As a first application of the formulas of Chap. 4, we recall the calculation of the vacuum energy density (the Casimir energy) corresponding to a massless scalar quantum field living in a compact universe, e.g. a Riemann sphere [126]. The zeta function regularization procedure supplemented with binomial expansion is a rigorous and well suited method for performing such analysis, as compared with other more involved techniques. The principal-part prescription (as described in Chap. 5) will be used to deal with the poles that eventually appear.

The investigation of the Casimir energy is one of the most basic issues of quantum field theory. But it turns out that the calculation for general configurations with curved boundaries is quite tricky, plagued with infinities and needing appropriate regularization. Sometimes cut-offs remain and it is difficult to extract uncontroverted, physically meaningful results [42]. This is why to have good numerical control on the functions that appear (along the line of the preceding sections) is such a basic issue. From a more practical point of view, the presence of the Casimir force in very different phenomena of condensed matter, solid state and laser physics has been established without any doubt, both theoretically and experimentally [156–162], so that the physical values to be matched are quite well known.

On the other hand, some of the most popular models of spacetime nowadays are given by manifolds without boundaries. Riemann spheres are to be counted among the simplest and most important of these manifolds. When one tries to calculate the determinants of the Laplacian operators on Riemann spheres by using zeta functions, one has to analytically continue expressions of the general form

$$f(s; a, b, c) \equiv \sum_{l=1}^{\infty} l^{-s+b} (l+a)^{-s+c}, \quad a > 0. \tag{6.1}$$

In [163, 164], for

$$\zeta^{(n)}(s) \equiv \sum_{l=1}^{\infty} \left[l^{-s}(l+2n+1)^{1-s} + l^{1-s}(l+2n+1)^{-s} \right], \tag{6.2}$$

the following result was obtained

$$\frac{d}{ds} \zeta^{(n)}(s) \bigg|_{s=0} = 4\zeta'(-1) - \frac{1}{2}(2n+1)^2 + \sum_{k=1}^{2n+1} (2k - 2n - 1) \ln k. \tag{6.3}$$

From here, in particular, for the zeta function of the Laplacian on the hemisphere with Dirichlet and Neumann boundary conditions, respectively, for which

$$\zeta_D(s) = \sum_{l=1}^{\infty} l[l(l+1)]^{-s},$$

$$\zeta_N(s) = \sum_{l=1}^{\infty} (l+1)[l(l+1)]^{-s}, \tag{6.4}$$

one gets

$$\zeta_D'(0) = 2\zeta'(-1) + \frac{1}{2}\ln(2\pi) - \frac{1}{4},$$

$$\zeta_N'(0) = 2\zeta'(-1) - \frac{1}{2}\ln(2\pi) - \frac{1}{4}. \tag{6.5}$$

Binomial expansion yields the following result:

$$f(s; a, b, c) \equiv \sum_{l=1}^{\infty} l^{-s+b}(l+a)^{-s+c} = \sum_{l=1}^{\infty} l^{-2s+b+c}\left(1 + al^{-1}\right)^{-s+c}$$

$$= \sum_{l=1}^{[a]} \{\,\} + \sum_{l=[a]+1}^{\infty} \{\,\}$$

$$= g(s; a, b, c) + \sum_{l=[a]+1}^{\infty} \sum_{k=0}^{\infty} \frac{\Gamma(1-s+c)}{k!\,\Gamma(1-s-k+c)} a^k l^{-2s-k+b+c}, \tag{6.6}$$

being $[a]$ the integer part of a, so that $g(s; a, b, c)$ is an analytic function of s, while the second, truncated series is absolutely convergent (since $al^{-1} < 1$ there). The final result is

$$f(s; a, b, c) = \sum_{l=1}^{[a]} l^{-2s+b+c}\left(1 + al^{-1}\right)^{-s+c} + \sum_{k=0}^{\infty} \frac{\Gamma(1-s+c)}{k!\,\Gamma(1-s-k+c)} a^k$$

$$\cdot \left[\zeta(2s+k-b-c) - \sum_{l=1}^{[a]} l^{-2s-k+b+c} \right]. \tag{6.7}$$

In particular,

$$f(0; a, b, c) = \sum_{l=1}^{[a]} l^{b+c}\left(1 + al^{-1}\right)^{c} + \sum_{k=0}^{\infty} \frac{\Gamma(1+c)}{k!\,\Gamma(1-k+c)} a^k$$

$$\cdot \left[\zeta(k-b-c) - \sum_{l=1}^{[a]} l^{-k+b+c} \right]. \tag{6.8}$$

and

$$f'(0; a, b, c) = -\sum_{l=1}^{[a]} \left[2\ln l + \ln\left(1 + al^{-1}\right)\right] l^{b+c}\left(1 + al^{-1}\right)^c$$

$$+ \sum_{k=0}^{\infty} \left\{ \frac{\psi(1 - k + c) - \psi(c)}{\Gamma(1 - k + c)} \left[\zeta(k - b - c) - \sum_{l=1}^{[a]} l^{-k+b+c}\right] \right.$$

$$\left. + \frac{2\Gamma(1 + c)}{\Gamma(1 - k + c)} \left[\zeta'(k - b - c) + \sum_{l=1}^{[a]} l^{-k+b+c} \ln l\right] \right\} \frac{a^k}{k!}. \quad (6.9)$$

In fact this last formula is a bit tricky, and has to be modified (in general) when b and c are integers. Then, the derivative for the particular value $k = b + 1$ must be taken with special care, by performing first expansions around the poles and zeros of these functions at $s = 0$.

For the zeta function $\zeta_D(s)$, (6.4), we obtain

$$\zeta_D(s) = \sum_{l=1}^{\infty} l^{1-2s}\left(1 + l^{-1}\right)^{-s}$$

$$= 2^{-s} + \sum_{k=0}^{\infty} \frac{\Gamma(1 - s)}{k!\Gamma(1 - s - k)}\left[\zeta(2s + k - 1) - 1\right] \quad (6.10)$$

and

$$\zeta_D'(0) = 2\zeta'(-1) + \frac{5}{4} + \frac{\gamma}{2} - \ln 2 - \sum_{k=2}^{\infty}(-1)^k\frac{\zeta(k) - 1}{k + 1}. \quad (6.11)$$

A second particular example is the following. For the case of a rectangle (of sides a and b) with Dirichlet boundary conditions, the spectrum of the Laplacian is $\lambda_{mn} = \pi^2(m^2/a^2 + n^2/b^2)$, and the zeta function

$$\zeta_{rec}(s) = \pi^{-2s} \sum_{m,n=1}^{\infty} \left(\frac{m^2}{a^2} + \frac{n^2}{b^2}\right)^{-s}$$

$$= -\frac{1}{2}\left(\frac{b}{\pi}\right)^{2s}\zeta(2s) + \frac{a}{2\sqrt{\pi}}\left(\frac{b}{\pi}\right)^{2s-1}\frac{\Gamma(s - 1/2)}{\Gamma(s)}\zeta(2s - 1)$$

$$+ \frac{2}{\Gamma(s)}\left(\frac{ab}{\pi}\right)^s\sqrt{\frac{a}{b}} \sum_{m,n=1}^{\infty} \left(\frac{m}{n}\right)^{s-1/2} K_{s-1/2}(2\pi mna/b), \quad (6.12)$$

where, once more, the corresponding asymptotic expansion for the Epstein zeta function in (4.49) has been used. Taking now the derivative, we get

$$\zeta'_{rec}(0) = \frac{1}{2}\ln(2b) + \frac{\pi a}{12b} + 2\sqrt{\frac{a}{b}}\sum_{m,n=1}^{\infty}\sqrt{\frac{n}{m}}K_{-1/2}(2\pi mna/b), \qquad (6.13)$$

which is best for numerical computations when $a \geq b$. (This is no restriction, indeed; however, one could also think of using (4.52) directly.) In the particular case $a = b$ (a square domain), it reduces to

$$\zeta'_{sq}(0) = \frac{1}{2}\ln(2a) + \frac{\pi}{12} + 2\sum_{m,n=1}^{\infty}\sqrt{\frac{n}{m}}K_{-1/2}(2\pi mn), \qquad (6.14)$$

which is just another expression for the same result obtained in [165]

$$\zeta'_{sq}(0) = \frac{1}{2}\ln(2a) + \frac{1}{4}\ln(8\pi) + \frac{1}{2}\ln\frac{\Gamma(3/4)}{\Gamma(1/4)}. \qquad (6.15)$$

The numerical value is, in both cases,

$$\zeta'_{sq}(0) = \frac{1}{2}\ln(2a) + 0.263672. \qquad (6.16)$$

Notice, however, that for the general rectangle, expression (6.13) is of much more practical use than the well-known one in terms of Dedekind's modular form η [165, 166]

$$\zeta'_{rec}(0) = \frac{1}{4}\ln(ab) - \ln\left[\frac{1}{\sqrt{2}}\left(\frac{b}{a}\right)^{1/4}\eta(q)\right],$$

$$\eta(q) = q^{1/24}\prod_{m=1}^{\infty}(1 - q^m), \quad q = \exp\left(-2\pi\sqrt{\frac{b}{a}}\right). \qquad (6.17)$$

In fact, a few first terms of the series in (6.13) suffice to obtain accurate numerical results (just as for the case of the square domain).

The three-dimensional Riemann sphere is a manifold without boundary that could perfectly well correspond to the spatial part of our universe, as a whole. The eigenvalues of the Laplacian operator are $\lambda_i^2 = l(l+2)/r^4$, with degeneracies $m_i = (l+1)^2$. Thus, the vacuum energy density for a massless scalar field is given by

$$E_3 = -\frac{\hbar}{2r^4}\zeta_3(s = -1/2), \quad \zeta_3(s) = \sum_{l=1}^{\infty}(l+1)^2[l(l+2)]^{-s}. \qquad (6.18)$$

We can write,

$$\zeta_3(s) = \sum_{l=2}^{\infty} l^{2(1-s)} \left(1 - l^{-2}\right)^{-s}$$

$$= \sum_{k=0}^{\infty} \frac{(-1)^k \Gamma(1-s)}{k! \Gamma(1-s-k)} \left[\zeta(2s+2k-2) - 1\right], \qquad (6.19)$$

and

$$\zeta_3(-1/2) = \sum_{k=0}^{\infty} \frac{(-1)^k \Gamma(3/2)}{k! \Gamma(3/2-k)} \left[\zeta(2k-3) - 1\right]$$

$$= -\frac{1}{16(s+1/2)} - 0.411502, \qquad (6.20)$$

which has a pole, for $k = 2$. Using the principal part prescription, we obtain

$$E_3 = -0.205751 \cdot \frac{\hbar}{r^4}. \qquad (6.21)$$

For the four-dimensional Riemann sphere, the corresponding eigenvalues and multiplicities are $\lambda_i^2 = l(l+3)/r^6$ and $m_i = (l+1)(l+2)(2l+3)/6$. The vacuum energy density is now

$$E_4 = -\frac{\hbar}{2r^5} \zeta_4(s=-1/2), \quad \zeta_4(s) = \frac{1}{6} \sum_{l=1}^{\infty} (l+1)(l+2)(2l+3) \left[l(l+3)\right]^{-s},$$
$$(6.22)$$

and we can write

$$\zeta_4(s) = \frac{1}{3} \sum_{l=1}^{\infty} u \left(u^2 - 1/4\right) \left(u^2 - 9/4\right)^{-s}$$

$$= \frac{1}{3} \sum_{k=0}^{\infty} \frac{(-1)^k \Gamma(1-s)}{k! \Gamma(1-s-k)} \left(\frac{9}{4}\right)^k$$

$$\cdot \left[\zeta(2s+2k-3, 5/2) - \zeta(2s+2k-1, 5/2)/4\right], \qquad (6.23)$$

being $u = l + 3/2$ and $\zeta(s, a)$ Hurwitz's zeta function

$$\zeta(s, a) = \sum_{n=0}^{\infty} (n+a)^{-s}. \qquad (6.24)$$

Again, nothing else has been done here but to apply the procedure as described above. We thus obtain

$$\zeta_4(-1/2) = \frac{1}{3} \sum_{k=0}^{\infty} \frac{(-1)^k \Gamma(3/2)}{k! \Gamma(3/2 - k)} \left(\frac{9}{4}\right)^k \left[\zeta(2k - 4, 5/2) - \zeta(2k - 2, 5/2)/4\right]$$
$$= -0.424550. \tag{6.25}$$

It is immediate to check that the term $(9/4)^k$ does not spoil convergence: we get a quickly convergent series, and finally

$$E_4 = -0.212275 \cdot \frac{\hbar}{r^5}. \tag{6.26}$$

An alternative treatment of the zeta functions above consists simply in splitting the polynomial in powers of the summation indices and then using the method of [163, 164]. It is just a matter of trying it to realize that that procedure is more lengthy than the one developed here. On the other hand, the cancellation of poles in this method must be done explicitly (resorting to expansions around all poles and zeros), while it is immediate in our procedure (actually, no pole is ever formed). We conclude that the most direct way turns out to be here both more rigorous and better suited for numerical evaluation. In connection with the general theory, as described in Chap. 4, we should observe that this is an example of the systematic use one can do of binomial expansion, i.e. the second of the two alternative expressions presented there. In the next section we shall describe an application of the first procedure, which stems also from the asymptotic expansion (4.50) in Chap. 4.

6.2 Application: Kaluza–Klein Model with Spherical Compactification

In [127], possible experimental manifestations of the contribution of heavy Kaluza–Klein particles, within a simple scalar model in six dimensions with spherical compactification, have been studied. The approach is based on the assumption that the inverse radius L^{-1} of the space of extra dimensions is of the order of the scale of the supersymmetry breaking $M_{SUSY} \sim 1 \div 10$ TeV. The total cross section of the scattering of two light particles has been calculated to one loop order and the effect of the Kaluza–Klein tower has been shown to be noticeable for energies $\sqrt{s} \geq 1.4L^{-1}$. The aim was to find an effect which could be measured experimentally and, with L^{-1} of the order of a few TeV, could actually be observed in future experiments.

By doing mode expansion, a multidimensional model on the spacetime $M^4 \times K$ (K is a compact manifold) can be represented as an effective theory on M^4 with an infinite set of particles, which is often referred to as the Kaluza–Klein tower of particles or modes (usually called *pyrgons*). The spectrum of the four-dimensional theory depends on the topology and geometry of K. The sector of the lowest state (of the zero mode) describes light particles (in the sense that their masses do not depend on L^{-1}) and coincides with the dimensionally reduced theory. Higher modes correspond to heavy particles with masses $\sim L^{-1}$. It is the contribution of pyrgons to physical quantities that might give evidence about the existence of extra dimensions. For simplicity, the scalar ϕ^4-model has been considered [127] and the space of extra dimensions is the two-dimensional sphere $\mathbb{S}^2$ of radius L, with an $SO(3)$-invariant metric on it. Study of quantum effects on spheres (mainly calculations of the Casimir effect and of the effective potential) can be found, for example, in [167–171]. The physical quantity to be calculated is the total cross section for the 2 *light particles* $\longrightarrow$ 2 *light particles* scattering process.

The main part of the calculation turns out to be a direct application of the results of Chap. 4. In fact, zeta function regularization techniques—based on the use of the first of the expansions considered there for the function $F(s; a, b^2)$ and its derivatives—provide the best way of treating sums over Kaluza–Klein modes which appear in the theory. Though there is some literature on the technique of performing calculations on spheres (see, for example, [35, 167–171] and references therein), the method used here is new and provides a simple and controllable way (from the numerical point of view) to deal with spherical compactification. Starting point of the analysis is the expression for the sum of the standard one-loop integrals over all Kaluza–Klein modes:

$$\Delta I\left(\frac{p^2}{M^2}, \frac{m_0}{M}, \epsilon\right) = i\pi^{2-\epsilon}\Gamma(\epsilon)\int_0^1 dx \sum_{l,m}' \left(\frac{M^2}{M_l^2}\right)^\epsilon \left[1 - \frac{p^2 x(1-x)}{M_l^2}\right]^{-\epsilon}, \quad (6.27)$$

where the prime means that the term for $l = 0$ is absent. Two limits, the low and high momentum expansion, respectively, can be considered:

$$(a) \quad (1-y)^{-\epsilon} = \sum_{k=0}^\infty \frac{\Gamma(k+\epsilon)}{k!\Gamma(\epsilon)} y^k, \quad y \equiv \frac{p^2 x(1-x)}{M_{l,m}^2}, \quad (6.28)$$

$$(b) \quad (1-y)^{-\epsilon} = (-y)^{-\epsilon}\left(1 - y^{-1}\right)^{-\epsilon}$$

$$= (-1)^{-\epsilon}\sum_{k=0}^\infty \frac{\Gamma(k+\epsilon)}{k!\Gamma(\epsilon)} y^{-k-\epsilon}. \quad (6.29)$$

There is no problem in doing the small-momentum expansion (a), which is valid for $|y| < 1$. In fact, since the maximum of $x(1-x)$ when $0 \le x \le 1$ is attained at $1/4$ (for $x = 1/2$), this formula is valid whenever $p^2 < 4M_{l,m}^2$. Nevertheless, the 'high-momentum expansion' (b) is much more difficult to perform. Actually, it is not possible to express its range of validity, $|y| < 1$, in terms of a simple inequality involving p^2 and $M_{l,m}^2$. As it stands, (6.29) is useless: we must first integrate (6.27) over x, in order to get rid of this unwanted dependence and then the formula yielding the desired expansion for $p^2 \ge 4M_{l,m}^2$ is different, according to different ranges of variation of p^2 in terms of $M_{l,m}^2$. The infinite sum over l gives rise to a derivative of the zeta-function $F(s; a, b^2)$ considered in Chap. 4. The x-integral yields just beta function factors.

The sums and integrals involved in the low-momentum expansion of (6.27) can be performed in the following order:

$$\Delta I\left(\frac{p^2}{M^2}, \frac{m_0}{M}, \epsilon\right) = i\pi^{2-\epsilon} \sum_{k=0}^{\infty} \frac{\Gamma(k+\epsilon)}{k!} B(k+1, k+1) \left(\frac{p^2}{M^2}\right)^k S_{k-1+\epsilon}, \quad (6.30)$$

where we have defined

$$S_{k+\epsilon} \equiv \sum_{l,m}' \left(\frac{M_{l,m}^2}{M^2}\right)^{-k-\epsilon} \quad (6.31)$$

and used

$$\int_0^1 dx \left[x(1-x)\right]^s = B(s+1, s+1), \quad (6.32)$$

$B(s, t) = \Gamma(s)\Gamma(t)/\Gamma(s+t)$ being Euler's beta function. In our particular case (spherical compactification) this sum reads

$$S_{k+\epsilon} = \sum_{l=1}^{\infty} (2l+1) \left[l(l+1) + \frac{m_0^2}{M^2}\right]^{-k-\epsilon}$$

$$= 2 \sum_{l=1}^{\infty} (l+1/2) \left[(l+1/2)^2 + \left(\frac{m_0^2}{M^2} - \frac{1}{4}\right)\right]^{-k-\epsilon}, \quad (6.33)$$

and can be written exactly as

$$S_{k+\epsilon} = \frac{1}{1-k-\epsilon} \frac{\partial}{\partial a} F(s-1; a, b)\Big|_{s=k+\epsilon, a=\frac{1}{2}, b=\frac{m^2}{M^2}-\frac{1}{4}}. \quad (6.34)$$

We will obtain explicitly the optimal truncation of this series, which has been numerically studied in Chap. 4. Taking these results into account, (6.34) yields

$$S_{k+\epsilon} = \sum_{n=0}^{\infty} \frac{(-1)^{n-1}}{n!} \frac{\Gamma(n+k+\epsilon-1)}{\Gamma(k+\epsilon)} b^{1-n-k-\epsilon} \tilde{\zeta}(-2n, 1/2), \quad (6.35)$$

where $\tilde{\zeta}$ has been defined in (4.54). For the first four terms of $S_{k+\epsilon}$ (providing the optimal truncation of the asymptotic series), we obtain

$$S_{k+\epsilon} = b^{1-k-\epsilon} \left[\frac{1}{k+\epsilon-1} + 2b^{-1}\zeta(-1, 1/2) - 2(k+\epsilon)b^{-2}\zeta(-3, 1/2)\right.$$

$$\left. + (k+\epsilon)(k+1+\epsilon)b^{-3}\zeta(-5, 1/2) - \cdots\right], \quad (6.36)$$

and, putting everything together, we arrive at

$$\Delta I\left(\frac{p^2}{M^2}, \frac{m_0}{M}, \epsilon\right)$$

$$= i\pi^{2-\epsilon}\Gamma(\epsilon)\left\{\left[\frac{b^{1-\epsilon}}{\epsilon-1} + 2b^{-\epsilon}\zeta(-1, 1/2) - 2\epsilon b^{-1-\epsilon}\zeta(-3, 1/2)\right.\right.$$

$$+ \epsilon(1+\epsilon)b^{-2-\epsilon}\zeta(-5, 1/2) - \frac{\epsilon(1+\epsilon)(2+\epsilon)}{3}b^{-3-\epsilon}\zeta(-7, 1/2) + \cdots\Bigg]$$

$$+ \frac{p^2}{M^2}B(2,2)\big[b^{-\epsilon} + 2\epsilon b^{-1-\epsilon}\zeta(-1, 1/2) - 2\epsilon(1+\epsilon)b^{-2-\epsilon}\zeta(-3, 1/2)$$

$$+ \epsilon(1+\epsilon)(2+\epsilon)b^{-3-\epsilon}\zeta(-5, 1/2) + \cdots\big]$$

$$+ \left(\frac{p^2}{M^2}\right)^2 B(3,3)\left[\frac{\epsilon b^{-1-\epsilon}}{2} + \epsilon(1+\epsilon)b^{-2-\epsilon}\zeta(-1, 1/2)\right.$$

$$- \epsilon(1+\epsilon)(2+\epsilon)b^{-3-\epsilon}\zeta(-3, 1/2)$$

$$\left.\left.+ \frac{\epsilon(1+\epsilon)(2+\epsilon)(3+\epsilon)}{2}b^{-4-\epsilon}\zeta(-5, 1/2) + \cdots\right] + \cdots\right\}. \qquad (6.37)$$

The numerical values of the coefficients are

$$B(1,1) = 1, \qquad B(2,2) = \frac{1}{6}, \qquad B(3,3) = \frac{1}{30}, \qquad \cdots, \qquad \frac{B(n+1, n+1)}{B(n,n)} \sim \frac{1}{4}, \qquad (6.38)$$

and

$$\zeta(-1, 1/2) = \frac{1}{24}, \qquad \zeta(-3, 1/2) = -\frac{7}{960}, \qquad \zeta(-5, 1/2) = \frac{31}{8064},$$

$$\zeta(-7, 1/2) = -\frac{127}{30720}, \qquad \zeta(-9, 1/2) = -\frac{511}{67584}, \qquad \cdots. \qquad (6.39)$$

We check that, in fact, the optimal truncation for the asymptotic series is obtained after the term $\zeta(-5, 1/2) = 0.00384$. As is apparent also, the resulting regularized series will not be valid for very small values of b. In that case, the second of the expressions of Chap. 4 has to be used, namely binomial expansion [127]. On the contrary, for large values of m_0/M the above series (6.37) is the one to be employed.

In a physical setting however, usually $m_0^2 \ll M^2$, so that $b \simeq -1/4$. One can also obtain convergent series for m_0/M small (even $m_0 = 0$ will be allowed) and valid for any finite p^2. To this end we must first perform the ϵ-expansion and then integrate over the x-variable:

$$I\left(\frac{p^2}{M^2}, \epsilon\right) = i\pi^{2-\epsilon}\Gamma(\epsilon) - i\pi^2 J\left(\frac{p^2}{4M^2}\right) + \mathcal{O}(\epsilon), \qquad (6.40)$$

where $J(z)$ is the finite part of the one-loop integral [172, 173]

$$J(z) = \int_0^1 dx \, \ln[1 - 4zx(1-x)], \qquad (6.41)$$

which is equal to

$$J(z) = \begin{cases} J_1(z) = 2\sqrt{\frac{z-1}{z}} \ln(\sqrt{1-z} + \sqrt{-z}) - 2, & \text{for } z \leq 0, \\[2mm] J_2(z) = 2\sqrt{\frac{1-z}{z}} \arctan\sqrt{\frac{z}{1-z}} - 2, & \text{for } 0 < z \leq 1, \\[2mm] J_3(z) = -i\pi\sqrt{\frac{z-1}{z}} + 2\sqrt{\frac{z-1}{z}} \ln(\sqrt{z} + \sqrt{z-1}) - 2, & \text{for } z > 1. \end{cases}$$
$$(6.42)$$

Now we can proceed with the summation over l, m. Taking the degeneracy in (6.27) into account and using ζ-regularization for the sums, we get

$$\Delta I\left(\frac{p^2}{M^2}, \frac{m_0}{M}, \epsilon\right) = i\pi^{2-\epsilon}\Gamma(\epsilon)\left[2\zeta(-1, 1/2) - 1\right]$$

$$- 2i\pi^2 \sum_{l=1}^{\infty}(l + 1/2)\ln\left[(l + 1/2)^2 + b\right]$$

$$- 2i\pi^2 \Delta J\left(\frac{p^2}{4M^2}; \frac{m_0}{M}\right) + \mathcal{O}(\epsilon). \qquad (6.43)$$

In particular, for $p^2 < 0$:

$$\Delta J\left(\frac{p^2}{4M^2}; \frac{m_0}{M}\right) = \sum_{l=1}^{\infty}(l + 1/2)J_1\left(\frac{p^2}{4M_l^2}\right), \qquad (6.44)$$

while for $p^2 > 0$:

$$\Delta J\left(\frac{p^2}{4M^2}; \frac{m_0}{M}\right) = \sum_{l=1}^{l^*(p)}(l + 1/2)J_3\left(\frac{p^2}{4M_l^2}\right) + \sum_{l=l^*(p)+1}^{\infty}(l + 1/2)J_2\left(\frac{p^2}{4M_l^2}\right), \qquad (6.45)$$

which contains, in general, an imaginary part. Here $l^*(p)$ is the maximum value of l which satisfies the inequality $4M^2(l + 1/2)^2 < p^2 - 4m_0^2$. If such l does not exist or is smaller than 1, we put $l^*(p) = 0$ and the first sum in (6.45) will be absent.

As we have already mentioned, the divergent sums over l are understood as being regularized by using the zeta-function regularization procedure. The calculation is then carried out in connection with this method. After expanding the functions under the summation signs in powers of $u_l = p^2/(4M_l^2)$ we are faced up with summations over the l-index, which give rise to Hurwitz zeta functions. As we clearly see from expression (6.43) above, the number of terms contributing to each sum changes with p. Thus, different explicit series are obtained for the different ranges of M^2/p^2. The first range, $|p^2/4M_1^2| < 1$, is somewhat special and deserves a careful treatment. According to the preceding analysis, only contributions in terms of a power series of $p^2/(4M^2)$ arise in this case, and we arrive to a series expansion which is the alternative to the low-momentum series that was obtained before, (6.37), now for small values of m_0^2/M^2, including the case $m_0^2 = 0$, i.e.

$$\Delta I\left(\frac{p^2}{M^2}, \frac{m_0}{M}, \epsilon\right) = i\pi^{2-\epsilon}\Gamma(\epsilon)\left[2\zeta(-1, 1/2) - 1\right]$$

$$- 2i\pi^2 \sum_{l=1}^{\infty}(l+1/2)\ln\left[(l+1/2)^2 + b\right]$$

$$+ 4i\pi^2 \sum_{l=1}^{\infty}(l+1/2)\left(\frac{u_l}{3} + \frac{2u_l^2}{15} + \frac{8u_l^3}{105}\right.$$

$$\left. + \frac{16u_l^4}{315} + \frac{128u_l^5}{3465} + \frac{256u_l^6}{9009} + \cdots\right) + \mathcal{O}(\epsilon),$$

$$u_l \equiv \frac{p^2}{4M_l^2} \equiv \frac{\frac{p^2}{4M^2}}{(l+1/2)^2 + b}, \quad b \equiv \frac{m_0^2}{M^2} - \frac{1}{4},$$

$$|p^2| < 4m_0^2 + 8M^2. \tag{6.46}$$

The l-sums yield again Epstein–Hurwitz zeta functions. In particular,

$$2\sum_{l=1}^{\infty}(l+1/2)u_l^k = \frac{1}{2(1-k)}\frac{\partial}{\partial a}F^{(1)}(k; a, b)\bigg|_{a=1/2}\left(\frac{p^2}{4M^2}\right)^k, \tag{6.47}$$

where a superindex (j) means 'j-truncated'—in the sense that the j first terms in the definitions of these zeta functions (for $n = 0, 1, \ldots, j-1$) are absent—namely

$$F^{(j)}(k; a, b) = \sum_{n=j}^{\infty}\left[(n+a)^2 + b\right]^{-k}, \quad j = 1, 2, \ldots \tag{6.48}$$

in terms of the function F introduced before. To finish, we shall now proceed with the evaluation of these derivatives that must be then substituted in (6.46), which provides the one-loop result for the sum over the Kaluza–Klein modes. Actually, it is not easy to obtain accurate numerical values in a simple way, but we will find some upper and lower bounds that will suffice for our purposes here.

Let us call

$$h^{(1)}(k; 1/2, b) \equiv \frac{1}{1-k}\frac{\partial}{\partial a}F_1^{(1)}(k; a, b)\bigg|_{a=1/2}. \tag{6.49}$$

The following bounds for these coefficients of the power series expansion will be useful for the general proof of convergence of the series (6.46). First, let us introduce the constants

$$\alpha_k \equiv \sum_{l=1}^{\infty} (2l + 1)\big[l(l + 1)\big]^{-k} < 2\zeta(2k - 1). \qquad (6.50)$$

They have the value

$$\alpha_1 = 1.1544, \qquad \alpha_2 = 0.9996, \qquad \alpha_3 = 0.4041,$$
$$\alpha_4 = 0.1918, \qquad \alpha_5 = 0.0944, \qquad \alpha_6 = 0.0470, \qquad (6.51)$$
$$\alpha_7 = 0.0235, \qquad \alpha_8 = 0.0117, \qquad \alpha_9 = 0.0059, \quad \dots$$

and from

$$m_0 = 0, \quad M \neq 0 \text{ arbitrary}: \quad h^{(1)}(k; 1/2, b) = \alpha_k,$$

$$m_0 \neq 0 \quad \begin{cases} m_0^2 \leq M^2/4: \quad 2\zeta^{(1)}(2k - 1, 1/2) \leq h^{(1)}(k; 1/2, b) \leq \alpha_k, \\ m_0^2 \geq M^2/4: \quad h^{(1)}(k; 1/2, b) \leq 2\zeta^{(1)}(2k, 1/2), \end{cases} \qquad (6.52)$$

we obtain

$$h^{(1)}(1; 1/2, b)\big|_{b \geq 0} \leq 0.93 \leq h^{(1)}(1; 1/2, b)\big|_{-1/4 \leq b \leq 0} \leq 1.15,$$

$$h^{(1)}(2; 1/2, b)\big|_{b \geq 0} \leq 0.83 \leq h^{(1)}(2; 1/2, b)\big|_{-1/4 \leq b \leq 0} \leq 0.99,$$

$$\qquad\qquad (6.53)$$

$$h^{(1)}(3; 1/2, b)\big|_{b \geq 0} \leq 0.29 \leq h^{(1)}(3; 1/2, b)\big|_{-1/4 \leq b \leq 0} \leq 0.40,$$

$$h^{(1)}(4; 1/2, b)\big|_{b \geq 0} \leq 0.12 \leq h^{(1)}(4; 1/2, b)\big|_{-1/4 \leq b \leq 0} \leq 0.19.$$

In few words, we see that with increasing m_0 the value of the coefficients decreases, starting from reasonable values (the α_k) for $m_0 = 0$. This makes of (6.46) a convenient approach to obtain the cross section of the problem and be able to compare with experimental results (see [127] for more details).

6.3   Critical Behavior of a Field Theory at Non-zero Temperature

Consider the effective potential of a four-fermion theory, that has been studied among others, by Inagaki, Kouno and Muta [174]. The phase-transition structure of the theory is to be investigated, in terms of the dimension D of the space where we are working. D is one of the variables and the final aim would be to study in detail the full structure of four-fermion theories with varying temperature and chemical potential, for arbitrary D between $2 \leq D < 4$, by using the $1/N$ expansion method. One wants to see, in particular, if chiral symmetry is or not restored for sufficiently high values of the temperature or the chemical potential.

To leading order in $1/N$, the effective potential $V_0(\sigma)$

$$V_0(\sigma) = \frac{1}{2\lambda_0}\sigma^2 + i\ln\det\left(i\gamma_\mu\partial^\mu - \sigma\right) + \mathcal{O}(1/N) \tag{6.54}$$

is essentially the one that provides the form of the effective potential of the theory. After the usual manipulations involving renormalization and integration over the angular variables, one finds that this effective potential is given by the expression [174]

$$V^{\beta\mu}(\sigma) = \frac{\sigma^2}{2\lambda} - \frac{\sqrt{\pi}}{\beta(2\pi)^{D/2}}\Gamma\left(\frac{1-D}{2}\right)\sum_{n=-\infty}^{+\infty}\left\{\left[\left(\frac{2n+1}{\beta}\pi - i\mu\right)^2\right]^{(D-1)/2}\right.$$
$$\left. + \left[\left(\frac{2n+1}{\beta}\pi - i\mu\right)^2 + \sigma^2\right]^{(D-1)/2}\right\}. \tag{6.55}$$

By direct application of one of the basic equations derived in Chap. 4, (2.93), we obtain

$$\sum_{n=-\infty}^{+\infty}\left[\left(\frac{2n+1}{\beta}\pi - i\mu\right)^2 + \sigma^2\right]^{(D-1)/2}$$
$$= \frac{\beta}{2\sqrt{\pi}}\frac{\Gamma(-D/2)}{\Gamma((1-D)/2)}\sigma^D$$
$$+ \frac{\beta^{1-D/2}(2\sigma)^{D/2}}{\sqrt{\pi}\Gamma((1-D)/2)}\sum_{n=1}^{\infty}(-1)^n n^{-D/2}\left(e^{\beta\mu n} + e^{-\beta\mu n}\right)K_{-D/2}(\beta\sigma n), \tag{6.56}$$

and

$$\sum_{n=-\infty}^{+\infty}\left[\left(\frac{2n+1}{\beta}\pi - i\mu\right)^2\right]^{(D-1)/2}$$
$$= \left(\frac{2\pi}{\beta}\right)^{D-1}\left[\varsigma\left(1-D, \frac{1}{2} - i\frac{\beta\mu}{2\pi}\right)\right.$$
$$\left. + \varsigma\left(1-D, -\frac{1}{2} + i\frac{\beta\mu}{2\pi}\right) - \left(\frac{1}{2} - i\frac{\beta\mu}{2\pi}\right)^{D-1}\right]. \tag{6.57}$$

The effective potential becomes

$$V^{\beta\mu}(\sigma) = \frac{\sigma^2}{2\lambda} - \frac{\sqrt{\pi}}{\beta(2\pi)^{D/2}}\left\{\Gamma\left(\frac{1-D}{2}\right)\left(\frac{2\pi}{\beta}\right)^{D-1}\left[\varsigma\left(1-D, \frac{1}{2} - i\frac{\beta\mu}{2\pi}\right)\right.\right.$$
$$\left.\left. + \varsigma\left(1-D, -\frac{1}{2} + i\frac{\beta\mu}{2\pi}\right) - \left(\frac{1}{2} - i\frac{\beta\mu}{2\pi}\right)^{D-1}\right] + \frac{\beta}{2\sqrt{\pi}}\Gamma\left(\frac{-D}{2}\right)\sigma^D\right.$$

$$+ \frac{\beta^{1-D/2}(2\sigma)^{D/2}}{\sqrt{\pi}} \sum_{n=1}^{\infty} (-1)^n n^{-D/2} \left(e^{\beta \mu n} + e^{-\beta \mu n} \right) K_{D/2}(\beta \sigma n) \Bigg\}. $$

$$(6.58)$$

The equation for the extrema of the potential leads us to the critical points (phase transitions), i.e.

$$\frac{\partial V^{\beta \mu}(\sigma)}{\partial \sigma} = 0. \qquad (6.59)$$

This yields

$$\frac{\sigma}{\lambda} - (2\pi)^{-D/2} \Bigg[\frac{1}{2} \Gamma \left(\frac{-D}{2} \right) D \sigma^{D-1} + \beta^{-D/2} D (2\sigma)^{D/2-1} $$

$$\cdot \sum_{n=1}^{\infty} (-1)^n n^{-D/2} \left(e^{\beta \mu n} + e^{-\beta \mu n} \right) K_{D/2}(\beta \sigma n) $$

$$+ \beta^{1-D/2}(2\sigma)^{D/2} \sum_{n=1}^{\infty} (-1)^n n^{1-D/2} \left(e^{\beta \mu n} + e^{-\beta \mu n} \right) K'_{D/2}(\beta \sigma n) \Bigg] = 0. \quad (6.60)$$

The series are convergent when $|\mu| \le \sigma$, for any D. In the first approximation, we have just

$$\frac{\sigma}{\lambda} - \frac{\Gamma(\frac{-D}{2}) D}{2(2\pi)^{D/2}} \sigma^{D-1} = 0, \qquad (6.61)$$

that is

$$\sigma_0 = \left[-\frac{2(2\pi)^{D/2} \Gamma(D/2+1) \sin(D\pi/2)}{\pi D \lambda} \right]^{1/(D-2)}. \qquad (6.62)$$

In the particular case when the number of dimensions is $5/2$ (an intermediate situation), we obtain

$$\sigma_0|_{D=5/2} = \frac{\Gamma(1/4)^2 \sqrt{\pi}}{2^{5/2} \lambda^2} = \frac{4.1187}{\lambda^2}. \qquad (6.63)$$

The self-consistency condition is $\sigma_0 \ge |\mu|$, namely,

$$|\mu| \le \left[-\frac{2(2\pi)^{D/2} \Gamma(D/2+1) \sin(D\pi/2)}{\pi D \lambda} \right]^{1/(D-2)}. \qquad (6.64)$$

(Notice that, for $D = 5/2$, this means $|\mu| < 4.12$ for $\lambda = 1$.)

The second approximation yields, with $\sigma = \sigma_0 + \sigma_1$

$$\frac{\sigma_1}{\lambda} - \frac{D(D-1)\Gamma(-D/2)}{2(2\pi)^{D/2}} \sigma_0^{D-2} \sigma_1 + \frac{\sigma_0^{(D-3)/2}}{\sqrt{2}(\pi \beta)^{(D-1)/2}} \left(\frac{D}{2\beta} - \sigma_0 \right) e^{-\beta(\sigma_0 - |\mu|)} \simeq 0. $$

$$(6.65)$$

The consistency check to second order is $\sigma_1 \ll \sigma_0$. We have

$$\frac{\sigma_1}{\sigma_0} \simeq \frac{\lambda}{(2-D)\sqrt{2}} \left(1 - \frac{D}{2\beta\sigma_0}\right) \left(\frac{\sigma_0}{\pi\beta}\right)^{(D-1)/2} \frac{1}{\sigma_0} e^{-\beta(\sigma_0 - |\mu|)} \ll 1. \qquad (6.66)$$

This is satisfied at high temperature T, if $\lambda \sim \beta = (kT)^{-1}$. In particular, the exponential is then $\sim \exp(-2kT)$, which is good for the quick convergence of the series of Bessel functions. For the particular case $D = 5/2$, substituting σ_0 into the last expression, we get

$$\left.\frac{\sigma_1}{\sigma_0}\right|_{D=5/2} \simeq -\left(1 - \frac{5\sqrt{2/\pi}\lambda^2}{\Gamma(1/4)^2\beta}\right) \frac{\lambda^{3/2}}{(2\pi)^{7/8}\sqrt{\Gamma(1/4)}\beta^{3/4}} e^{-\beta(\sigma_0 - |\mu|)}$$

$$= -0.1052\left(1 - 0.3035\frac{\lambda^2}{\beta}\right) \frac{\lambda^{3/2}}{\beta^{3/4}} e^{-\beta(4.1187/\lambda^2 - |\mu|)}. \qquad (6.67)$$

This looks as a very reasonable dependence and it certainly allows us to proceed with the analytical treatment of the general situation, in a sort of perturbative way (order by order).

6.4 Application to Quantizing Through the Wheeler–De Witt Equation

In a work by S. Carlip [175], dealing with the approach to $(2+1)$-dimensional quantum gravity which consists in making direct use of the Wheeler–De Witt equation, this author came across a rather involved mathematical problem. (By the way, none of the approaches that have been employed for the quantization of gravity is simple, for different reasons.) As in the examples above, here we will concentrate only in the specific points of the whole question that have to do with the methods developed in the book. They concern the calculation of the determinant that appears in his method and has to be evaluated on the fundamental domain in two dimensions. It is the determinant of a differential operator, D_0, which has the following set of eigenfunctions and eigenvalues

$$|mn\rangle = e^{2\pi i (mx+ny)}, \qquad l_{mn} = \frac{4\pi^2}{\tau_2}|n - m\tau|^2 + V_0, \qquad (6.68)$$

where m and n are integers, and τ and τ_2 are the usual labels corresponding to the standard two-dimensional metric on the domain

$$d\bar{s}^2 = \tau_2^{-1}|dx + \tau\, dy|^2, \qquad (6.69)$$

with x and y angular coordinates of period 1 and $\tau = \tau_1 + i\tau_2$ the modulus (a complex parameter). V_0 is the spatial integral of the relevant potential function.

6.4.1 Explicit Zeta-Function Calculation of the Essential Determinant and Extrema of the Potential

At that point, the difficulty has boiled down to a well formulated mathematical problem which has a straightforward solution by means of the expressions derived in the preceding chapters, through the calculation of the corresponding zeta function, ζ_{D_0}. In fact, after simplifying the notation a little, we easily recognize that we have to deal here with a series of the form (7.18). For $s = 1$ (the value of interest) the analytic continuation hits a pole and, therefore, it must be conveniently defined [126].

The quantization of gravity in a $(2+1)$-dimensional spacetime—with the metric of the spatial part being given by (6.69)—proceeds through the calculation of the zeta function corresponding to the basic differential operator D_0, which has the spectral decomposition given above (6.68). In terms of the function $E(s; a, b, c; q)$, the zeta function of D_0 is

$$\zeta_{D_0}(s) = E\left(s; 4\pi^2/\tau_2, -8\pi^2\tau_1/\tau_2, 4\pi^2\left(\tau_1^2 + \tau_2^2\right)/\tau_2; V_0\right). \qquad (6.70)$$

In Fig. 7 we can see a plot of $E(s)$ for specified values of the parameters a, b, c and q in function of τ_1 and τ_2. One has, in particular, $\Delta = 64\pi^4$ and, using (4.32),

$$
\begin{aligned}
\zeta_{D_0}(s) = {}& \frac{2^{-2s+1}\pi^{-2s}}{\tau_2^{-s}} \zeta_{EH}\left(s, V_0\tau_2/\left(4\pi^2\right)\right) \\
& + \frac{2^{-2s+1}\pi^{-2s+1/2}\Gamma(s - 1/2)}{\tau_2^{s-1}\Gamma(s)} \zeta_{EH}\left(s - 1/2, V_0/\left(4\pi^2\tau_2\right)\right) \\
& + \frac{2^{-2s+2}\pi^{-s}\sqrt{\tau_2}}{\Gamma(s)} \sum_{n=0}^{\infty} n^{s-1/2}\cos(2n\pi\tau_1) \sum_{d|n} d^{1-2s} \int_0^{\infty} dt\, t^{s-3/2} \\
& \cdot \exp\left\{-n\pi\tau_2\left[\left(1 + \frac{V_0}{4\pi^2 d^2\tau_2}\right)t + t^{-1}\right]\right\}.
\end{aligned}
\qquad (6.71)
$$

The quantity of interest is the determinant of the operator D_0. This is most conveniently computed by means of its zeta function. In particular:

$$\det{}^{1/2} D_0 = \exp\left[-\frac{1}{2}\zeta'_{D_0}(0)\right]. \qquad (6.72)$$

Thus, we must now calculate the derivative of (4.32) at $s = 0$. We have, for the general function $E(s; a, b, c; q)$

$$
\begin{aligned}
& E'(0; a, b, c; q) \\
& = \ln a + 2\zeta'_{EH}(0, q/a) + \sigma'(0, q) \\
& \quad + 4\sum_{n=1}^{\infty} n^{-1}\cos(n\pi b/a) \sum_{d|n} d\exp\left[-\frac{\pi n}{a}\left(\Delta + \frac{4aq}{d^2}\right)^{1/2}\right],
\end{aligned}
\qquad (6.73)
$$

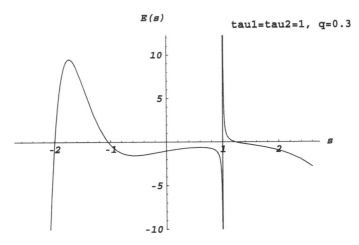

Fig. 7 Plot of the function of s corresponding to the analytical continuation of the inhomogeneous Epstein zeta function $E(s; a, b, c; q)$, as given by the extension of the Chowla–Selberg formula derived in Chap. 4. The values of the parameters have been fixed as indicated (in terms of τ_1 and τ_2)

where

$$\zeta'_{EH}(0; p) = -\pi \sqrt{p} + \frac{1}{2} \ln p - \ln\left(1 - e^{-2\pi \sqrt{p}}\right), \qquad (6.74)$$

and

$$\sigma'(0, q) = 2\pi \sqrt{\frac{q}{a}} + \frac{2\pi q}{\sqrt{\Delta}}(-1 + \ln q) + 4\sqrt{\frac{q}{a}} \sum_{n=1}^{\infty} n^{-1} K_1\left(4n\pi \sqrt{\frac{aq}{\Delta}}\right). \qquad (6.75)$$

We get

$$E'(0; a, b, c; q)$$

$$= -\frac{2\pi q}{\sqrt{\Delta}} + \left(1 + \frac{2\pi q}{\sqrt{\Delta}}\right) \ln q$$

$$- 2\ln\left(1 - e^{-2\pi \sqrt{q/a}}\right) + 4\sqrt{\frac{q}{a}} \sum_{n=1}^{\infty} n^{-1} K_1\left(4n\pi \sqrt{\frac{aq}{\Delta}}\right)$$

$$+ 4 \sum_{n=1}^{\infty} n^{-1} \cos(n\pi b/a) \sum_{d|n} d \exp\left[-\frac{\pi n}{a}\left(\Delta + \frac{4aq}{d^2}\right)^{1/2}\right]. \qquad (6.76)$$

Finally, for the determinant of D_0, we obtain

$$\det{}^{1/2} D_0 = \frac{1}{\sqrt{V_0}}\left(1 - e^{-\sqrt{V_0 \tau_2}}\right) \exp\left\{\frac{V_0}{8\pi}(1 - \ln V_0)\right.$$

$$-\frac{1}{\pi}\sqrt{\tau_2 V_0}\sum_{n=1}^{\infty}n^{-1}K_1\left(n\sqrt{\frac{V_0}{\tau_2}}\right)-2\sum_{n=1}^{\infty}n^{-1}\cos(2n\pi\tau_1)$$

$$\cdot\sum_{d|n}d\exp\left[-2n\pi\tau_2\left(1+\frac{V_0}{4\pi^2\tau_2 d^2}\right)^{1/2}\right]\Bigg\}. \tag{6.77}$$

We observe again that the final formula is really simple for practical purposes, since it provides a good approximations with just a few terms, which are, on its turn, elementary functions of the relevant variables and parameters. This is so, because the infinite series that appear converge quickly (terms are exponentially decreasing with n). In an asymptotical approach to the determinant, only the first terms in (6.77) are relevant.

From the detailed analysis in [175], it follows that the quantity to be calculated is the derivative with respect to V_0 of the above determinant, since this quantity vanishes precisely at the solutions of the Hamiltonian constraint (always in the language of quantization through the corresponding Wheeler–De Witt equation). In other words, the solutions of the equation

$$\frac{\partial}{\partial V_0}\det^{1/2}D_0=0, \tag{6.78}$$

will yield the conditions that the quantized magnitudes and parameters are bound to satisfy as a consequence of the Wheeler–De Witt equations. After some work, (6.78) can be written as

$$\det^{1/2}D_0\cdot\left[-\frac{1}{8\pi}\ln V_0-\frac{1}{2V_0}+\frac{1}{2}\sqrt{\frac{\tau_2}{V_0}}\left(e^{\sqrt{V_0\tau_2}}-1\right)^{-1}\right.$$

$$-\frac{1}{2\pi}\sqrt{\frac{\tau_2}{V_0}}\sum_{n=1}^{\infty}n^{-1}K_1\left(n\sqrt{\frac{V_0}{\tau_2}}\right)-\frac{1}{2\pi}\sum_{n=1}^{\infty}K_1'\left(n\sqrt{\frac{V_0}{\tau_2}}\right)$$

$$\left.+\frac{1}{2\pi}\sum_{n=1}^{\infty}\cos(2n\pi\tau_1)\sum_{d|n}\left(d^2+\frac{V_0}{4\pi^2\tau_2}\right)^{-1/2}\exp\left(-2n\pi\tau_2\sqrt{1+\frac{V_0}{4\pi^2\tau_2 d^2}}\right)\right]$$

$$=0, \tag{6.79}$$

where the primes mean derivatives of the Bessel functions. The discussion can be continued analytically in the limit $\tau_2\gg 1$ without further problem (for consistency it must be $V_0\tau_2\gg 1$ at the critical points V_0). Numerical plots of the functions above are given in the accompanying figures. A short summary of the numerical analysis for a sample of representative values of the parameters τ_1 and τ_2 can be read off from Figs. 8, 9, 10. Very interesting behaviors appear, which are completely different for the different ranges of values of the parameters. In principle—the consistency of the approximation is to be checked *a posteriori*—(6.79) can be reduced to the a simple expression which includes, at most, the first of the terms involving Bessel

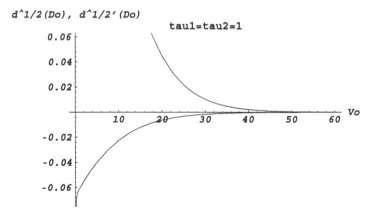

Fig. 8 Plot of the square root of the determinant $\det^{1/2} D_0$ and of its derivative as a function of V_0, for $\tau_1 = \tau_2 = 1$. No minimum of the determinant appears for these specific values of τ_1 and τ_2

functions (we can here employ the same treatment as described in detail for the previous application).

6.4.2 An Alternative Treatment by Means of Eisenstein Series

An alternative way of treating the general case is the following (see [175]). The inhomogeneity (the q term here) is taken care of by the simplest (but hardly economic) method of performing a binomial expansion of the sort [126]

$$\sum_{k=0}^{\infty} \frac{\Gamma(s+k)}{k!\Gamma(s)} q^k E(z, s+k), \tag{6.80}$$

where $E(z, s)$ is an Eisenstein series (see, for instance, Lang [176] or Kubota [177]), which is obtained from $F(s; a, b, c; 0)$ by doing the substitution

$$2z = a + iu, \qquad c = C\frac{u}{2}, \tag{6.81}$$

so that

$$E(z, s) = \sum_{m,n=0}^{\infty}{}' (u/2)^s |m + nz|^{-2s}, \tag{6.82}$$

and has the series expansion

$$E(z, s) = 2\zeta(2s) + 2\sqrt{\pi}(u/2)^{1-s} \frac{\Gamma(1 - s/2)\zeta(2s - 1)}{\Gamma(s)}$$

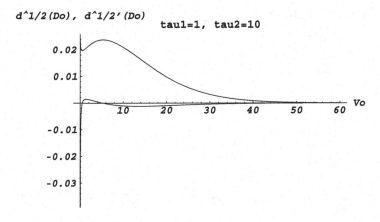

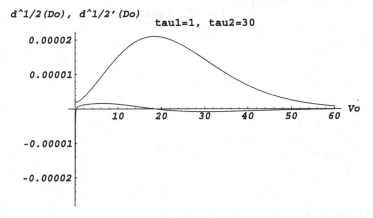

Fig. 9 *Top*: Plot of the square root of the determinant $\det^{1/2} D_0$ and of its derivative as a function of V_0, for $\tau_1 = 1$, $\tau_2 = 10$. Two extrema of the determinant appear for these values of τ_1 and τ_2. The roots of the derivatives are obtained for $V_0 = 0.48$ and $V_0 = 5.36$. *Bottom*: The same for $\tau_1 = 1$, $\tau_2 = 30$. The two extrema of the determinant have now moved to $V_0 = 0.31$ and $V_0 = 17.52$

$$+2\sum_{m=1}^{\infty}\sum_{n\neq 0}e^{i\pi mna}\left(\frac{2|n|}{mu}\right)^{s-1/2}K_{s-1/2}(\pi mnu/2). \qquad (6.83)$$

It is important to notice, however, that when doing things in this last way the final result is expressed in terms of three infinite sums, while in the first general procedure only one infinite sum appears (together with a finite sum, for every index n, over the divisors of n), and it is quickly convergent. Notice, moreover, how the d-term in the exponent in (4.32), when expanded in power series, gives rise to the binomial sum corresponding to the last treatment. The advantage of the use of the method developed here stemming from (4.32), seems clear (expanding a negative exponential is in general computationally disastrous).

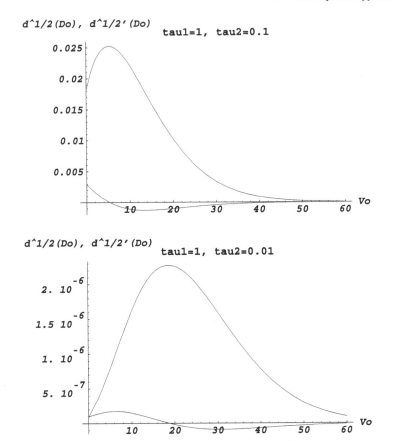

Fig. 10 *Top*: Plot of the square root of the determinant and of its derivative as a function of V_0, for $\tau_1 = 0.1$, $\tau_2 = 1$. Here only one extremum of the determinant appears, at $V_0 = 5.33$. *Bottom*: The same for $\tau_1 = 0.01$, $\tau_2 = 1$. The extremum persists and has now moved to $V_0 = 19.68$

To summarize this section, two problems were singularized out in [175] as the main difficulties that appear in the quantization of $(2 + 1)$-dimensional gravity through the Wheeler–De Witt equation:

1. To give grounds for the choice of the specific operator ordering of the Hamiltonian constraint which leads to the Wheeler–De Witt equation of the quantized system.
2. To understand the functional dependence of the determinant $\det^{1/2} D_0$ in terms of the relevant variables and to obtain its extrema as a function of the potential V_0.

With the use of zeta-function techniques, we are here able to solve the second one, by means of a completely consistent approximation. We should also mention that the advantages of the present method with respect to the use of binomial expansions and the Eisenstein series are quite conclusive.

Apart from the applications that we have here considered—directly dealing with the generalization of the Chowla–Selberg formula derived above—in general our new expression will be certainly useful in many situations involving massive theories, finite temperatures or a chemical potential in a compactified spacetime. This is the meaning to be attributed to the constant q in the more usual mathematical-physics problems. As for the possible mathematical applications, the introduction of the constant q into the Epstein function extends the scope of the applications already envisaged by Chowla and Selberg, in the sense that now one is perfectly able to develop a sort of classification of the problems according to the different values of q.

6.5   Spectral Zeta Function for Both Scalar and Vector Fields on a Spacetime with a Noncommutative Toroidal Part

We shall now consider the physical example of a quantum system consisting of scalars and vector fields on a D-dimensional noncommutative manifold, M, of the form $\mathbb{R}^{1,d} \bigotimes \mathbb{T}_\theta^p$ (thus $D = d + p + 1$). $\mathbb{T}_\theta^p$ is a p-dimensional noncommutative torus, its coordinates satisfying the usual relation: $[x_j, x_k] = i\theta\sigma_{jk}$. Here σ_{jk} is a real nonsingular, antisymmetric matrix of ± 1 entries, and θ is the noncommutative parameter.

This physical system has attracted much interest recently, in connection with M-theory and with string theory [178–184], and also because of the fact that those are perfectly consistent theories by themselves, which could lead to brand new physical situations. It has been shown, in particular, that noncommutative gauge theories describe the low energy excitations of open strings on D-branes in a background Neveu–Schwarz two-form field [178–180].

This interesting system provides us with a quite non-trivial case where the formulas derived above are indeed useful. For one, the zeta functions corresponding to bosonic and fermionic operators in this system are of a different kind, never considered before. And, moreover, they can be most conveniently written in terms of the zeta functions in Sect. 4.2. What is also nice is the fact that a unified treatment (with just *one* zeta function) can be given for both cases, the nature of the field appearing there as a simple parameter, together with those corresponding to the numbers of compactified, noncompactified, and noncommutative dimensions of the spacetime.

6.5.1 Poles of the Zeta Function

The spectral zeta function for the corresponding (pseudo-)differential operator can be written in the form [185]

$$\zeta_\alpha(s) = \frac{V}{(4\pi)^{(d+1)/2}} \frac{\Gamma(s - (d+1)/2)}{\Gamma(s)} {\sum_{\vec{n} \in \mathbb{Z}^p}}' Q(\vec{n})^{(d+1)/2-s}$$

$$\cdot \left[1 + \Lambda\theta^{2-2\alpha} Q(\vec{n})^{-\alpha} \right]^{(d+1)/2-s}, \tag{6.84}$$

where $V = \mathrm{Vol}(\mathbb{R}^{d+1})$, the volume of the non-compact part, and $Q(\vec{n}) = \sum_{j=1}^p a_j n_j^2$, a diagonal quadratic form, being the compactification radii $R_j = a_j^{-1/2}$. Moreover, the value of the parameter $\alpha = 2$ for scalar fields and $\alpha = 3$ for vectors, distinguishes between the two different fields. In the particular case when we set all the compactification radii equal to R, we obtain:

$$\zeta_\alpha(s) = \frac{V}{(4\pi)^{(d+1)/2}} \frac{\Gamma(s - (d+1)/2)}{\Gamma(s) R^{d+1-2s}} {\sum_{\vec{n} \in \mathbb{Z}^p}}' I(\vec{n})^{(d+1)/2-s}$$

$$\cdot \left[1 + \Lambda\theta^{2-2\alpha} R^{2\alpha} I(\vec{n})^{-\alpha} \right]^{(d+1)/2-s}, \tag{6.85}$$

being now the quadratic form: $I(\vec{n}) = \sum_{j=1}^p n_j^2$.

After some calculations, this zeta function can be written in terms of the Epstein zeta function of Sect. 4.2, with the result:

$$\zeta_\alpha(s) = \frac{V}{(4\pi)^{(d+1)/2}} \sum_{l=0}^\infty \frac{\Gamma(s + l - (d+1)/2)}{l!\Gamma(s)} \left(-\Lambda\theta^{2-2\alpha}\right)^l$$

$$\cdot \zeta_{Q,\vec{0},0}(s + \alpha l - (d+1)/2), \tag{6.86}$$

which reduces, in the particular case of equal radii, to

$$\zeta_\alpha(s) = \frac{V}{(4\pi)^{(d+1)/2} R^{d+1-2s}} \sum_{l=0}^\infty \frac{\Gamma(s + l - (d+1)/2)}{l!\Gamma(s)} \left(-\Lambda\theta^{2-2\alpha}\right)^l$$

$$\cdot \zeta_E(s + \alpha l - (d+1)/2), \tag{6.87}$$

where we use here the notation $\zeta_E(s) \equiv \zeta_{I,\vec{0},0}(s)$, e.g., the Epstein zeta function for the standard quadratic form.

The pole structure of the resulting zeta function deserves a careful analysis. It differs, in fact, very much from all cases that were known in the literature till now. This is not difficult to understand, from the fact that the pole of the Epstein zeta function at $s = p/2 - \alpha k + (d+1)/2 = D/2 - \alpha k$, when combined with the poles of the gamma functions, yields a very rich pattern of singularities for $\zeta_\alpha(s)$, on taking into account the different possible values of the parameters involved. The pole structure is straightforwardly found from the explicit expressions for the zeta functions in Sect. 4.2.

Having already given the formula (6.86) above—that contains everything needed to perform such calculation of pole position, residua and finite part—for its importance for the calculation of the determinant and the one-loop effective action from

Table 5 Pole structure of the zeta function $\zeta_\alpha(s)$, at $s = 0$, according to the different possible values of d and D ($\dot{\overline{2\alpha}}$ means *multiple* of 2α)

$$
\text{For } d = 2k \quad
\begin{cases}
\text{if } D \neq \dot{\overline{2\alpha}} & \Longrightarrow \quad \zeta_\alpha(0) = 0, \\
\text{if } D = \dot{\overline{2\alpha}} & \Longrightarrow \quad \zeta_\alpha(0) = \text{finite.}
\end{cases}
$$

$$
\text{For } d = 2k - 1 \quad
\begin{cases}
\text{if } D \neq \dot{\overline{2\alpha}} \; \begin{Bmatrix} \text{finite}, & \text{for } l \leq k \\ 0, & \text{for } l > k \end{Bmatrix} & \Longrightarrow \quad \zeta_\alpha(0) = \text{finite,} \\[2em]
\text{if } D = 2\alpha l \; \begin{Bmatrix} \text{pole}, & \text{for } l \leq k \\ \text{finite}, & \text{for } l > k \end{Bmatrix} & \Longrightarrow \quad \zeta_\alpha(0) = \text{pole.}
\end{cases}
$$

the zeta function, we will here start by specifying what happens at $s = 0$. Remarkably enough, a pole appears in many cases (depending on the values of the different parameters). This will also serve as an illustration of what one has to expect for other values of s. The general case will be left for the following subsection.

It is convenient to classify the different possible subcases according to the values of d and $D = d + p + 1$. We obtain, at $s = 0$, the pole structure given in Table 5.

Here l is the summation index in (6.86). The appearance of a pole of the zeta function $\zeta_\alpha(s)$, for both values of α, at $s = 0$ is, let us repeat, an absolute novelty, bound to have important physical consequences for the regularization process. It is necessary to observe, that this fact is *not* in contradiction with the well known theorems on the pole structure of a (elliptic) differential operator [186]. The situation that appears in the noncommutative case is completely different. (i) To begin with, we do not have any longer a standard differential operator, but a strictly *pseudo-differential* one, from the beginning. (ii) Moreover, the new spectrum is *not* perturbatively connected (for $\theta \to 0$) with the corresponding one for the commutative case.

6.5.2 Explicit Analytic Continuation of $\zeta_\alpha(s)$, $\alpha = 2, 3$, in the Complex s-Plane

Substituting the corresponding formula, from the preceding section, for the Epstein zeta functions in (6.86), we obtain the following explicit analytic continuation of $\zeta_\alpha(s)$ ($\alpha = 2, 3$), for bosonic and fermionic fields, to the *whole* complex s-plane:

$$
\zeta_\alpha(s) = \frac{2^{s-d} V}{(2\pi)^{(d+1)/2} \Gamma(s)} \sum_{l=0}^\infty \frac{\Gamma(s + l - (d+1)/2)}{l! \, \Gamma(s + \alpha l - (d+1)/2)} \left(-2^\alpha \Lambda \theta^{2 - 2\alpha} \right)^l
$$

$$
\cdot \sum_{j=0}^{p-1} (\det A_j)^{-1/2} \left[\pi^{j/2} a_{p-j}^{-s - \alpha l + (d+j+1)/2} \Gamma(s + \alpha l - (d+j+1)/2) \right.
$$

$$\cdot\, \zeta_R(2s + 2\alpha l - d - j - 1) + 4\pi^{s+\alpha l-(d+1)/2} a_{p-j}^{-(s+\alpha l)/2-(d+j+1)/4}$$

$$\cdot \sum_{n=1}^{\infty} \sum_{\vec{m}_j \in \mathbb{Z}^j}{}' n^{(d+j+1)/2-s-\alpha l} \left(\vec{m}_j^{\,t} A_j^{-1} \vec{m}_j \right)^{(s+\alpha l)/2-(d+j+1)/4}$$

$$\cdot\, K_{(d+j+1)/2-s-\alpha l} \left(2\pi n \sqrt{a_{p-j} \vec{m}_j^{\,t} A_j^{-1} \vec{m}_j} \,\right) \Bigg]. \tag{6.88}$$

As discussed in the previous subsection in detail, the non-spurious poles of this zeta function are to be found in the terms corresponding to $j = p - 1$. With the knowledge we have gained from the analytical continuation of the Epstein zeta functions in Sect. 6.2, the final analysis can be here completed at once. Note that the situation here corresponds to the diagonal case with $c_1 = \cdots = c_p = q = 0$.

To be remarked again is that, what we have in the end, by using our method, is an exponentially fast convergent series of Bessel functions together with a first, finite part, where a pole (simple or double, as we shall see) may show up, for specific values of the dimensions of the different parts of the manifold, depending also on the nature (scalar vs. vectorial) of the fields (the value of α, see Table 5 and (6.88)).

To summarize the discussion at the end of the preceding section, the pole structure of (6.88) is in fact best seen from (6.86) (for $s = 0$ it has been analyzed in Table 5 already). For a fixed value of the summation index l, the contribution to the only pole of the zeta function $\zeta_E(s + \alpha l - (d+1)/2)$, at $s = D/2 - \alpha l$, comes from the last term of the j-sum only, namely from $j = p - 1$. It is easy to check that it yields the corresponding residuum (4.46). This corresponds to the second sum in (6.88). Combined now with the poles of the gamma functions, and taking into account the first series in l, this yields the following expression for the residua of the zeta function $\zeta_\alpha(s)$ at the poles $s = D/2 - \alpha l, l = 0, 1, 2, \ldots$

$$\mathrm{Res}\, \zeta_\alpha(s) \Big|_{s=D/2-\alpha l} = \frac{2^{p/2-d} \pi^{(p-d-1)/2} V}{\Gamma(p/2)} (\det A_p)^{-1/2} \frac{(-\Lambda \theta^{2-2\alpha})^l}{l!},$$

$$\cdot\, \frac{\Gamma(p/2 + (1-\alpha)l)}{\Gamma(D/2 - \alpha l)}, \quad l = 0, 1, 2, \ldots. \tag{6.89}$$

Actually, depending on D and p being even or odd, completely different situations arise, for different values of l: from the disappearance of the pole, giving rise to a finite contribution, to the appearance of a simple or a double pole. We shall distinguish four different situations and, to simplify the notation, we will denote by U the whole factor in the expression (6.89) for the residuum, that multiplies the last fraction of two gamma functions (in short, $\mathrm{Res}\, \zeta_\alpha = U\Gamma_1/\Gamma_2$).

1. For $D - 2\alpha l = -2h, h = 0, -1, -2, \ldots$
   a. for $p/2 + (1 - \alpha)l, l \neq 0, -1, -2, \ldots \implies$ finite, $\mathrm{Res}\, \zeta_\alpha = -h! U \Gamma(p/2 + (1-\alpha)l)$;
   b. for $p = 2(\alpha - 1)l - 2k, k = 0, -1, -2, \ldots \implies$ pole, $\mathrm{Res}\, \zeta_\alpha = (h!/k!)U$.
2. For $D - 2\alpha l \neq -2h, h = 0, -1, -2, \ldots$

Table 6 General pole structure of the zeta function $\zeta_\alpha(s)$, according to the different possible values of D and p being odd or even. In italics, the type of behavior corresponding to lower values of l is quoted, while the behavior shown in roman characters corresponds to larger values of l

$p\backslash D$	Even	Odd
odd	(1a) *pole*/finite $(l \geq l_1)$	(2a) *pole*/pole
even	(1b) *double pole*/pole $(l \geq l_1, l_2)$	(2b) *pole*/double pole $(l \geq l_2)$

a. for $p/2 + (1 - \alpha)l,\ l \neq 0, -1, -2, \ldots \implies$ pole, Res $\zeta_\alpha = U\Gamma(p/2 + (1 - \alpha)l)/\Gamma(D/2 + \alpha l)$;

b. for $p = 2(\alpha - 1)l - 2k,\ k = 0, -1, -2, \ldots \implies$ double pole, Res $\zeta_\alpha = (-1/k!)U/\Gamma(D/2 + \alpha l)$.

Note that we here just quote the *generic* situation that occurs for l *large* enough in each case. For instance, if $p = 2$ a double pole appears for $l = 1, 2, \ldots$. For $p = 4$, a double pole appears for $l = 1, 2, \ldots$, if $\alpha = 3$, but only for $l = 2, 3, \ldots$, if $\alpha = 2$. For $p = 6$, a double pole appears for $l = 2, 3, \ldots$, if $\alpha = 3$, but only for $l = 3, 4, \ldots$, if $\alpha = 2$, and so on. The case with both D and p even (what implies d odd) is the most involved one. For $p = 2$ and $D = 4$, for instance, there is a transition from a pole for $l = 0$ corresponding to the zeta function factor, to a pole for $l = 1$ and higher, corresponding to the gamma function in the numerator (the compensation of the pole of the zeta function factor with the one coming from the gamma function in the denominator prevents the formation of a double pole). In any case, the explicit analytic continuation of $\zeta_\alpha(s)$ given by (6.88) contains *all* the information one needs for calculating the poles and corresponding residua in a straightforward way.

The pole structure can be summarized as in Table 6.

An application of these formulas to the calculation of the one-loop partition function corresponding to quantum fields at finite temperature, on a noncommutative flat spacetime, can be found in [187].

Chapter 7
Miscellaneous Applications Combining Zeta with Other Regularization Procedures

In this chapter the following applications of the method of zeta-function regularization will be described: (i) First, some aspects of the comparison that has been established recently by Fujikawa between the generalized Pauli–Villars regularization and the covariant regularization of composite current operators will be investigated. (ii) Second, a calculation of the Casimir energy for the transverse oscillations of a piecewise uniform closed string will be performed. The string consists of two parts, each having in general different tension and mass density but adjusted in such a way that the velocity of sound always equals the velocity of light. This model was introduced by I. Brevik and H.B. Nielsen. For the calculation, a nice modification of the method of the zeta function as described till now will be necessary, in the sense that it will be combined with some basic theorems of complex analysis. Also, a comparison with the results obtained by means of the introduction of a cutoff will be established which provides additional physical insight to the zeta function procedure. Hadamard regularization is also discussed, as a very useful auxiliary tool to the zeta method, in dealing with additional infinities and physical cut-offs. This aspect of comparing zeta-function analytic continuation with other regularization procedures is the common point of the examples studied here.

7.1 Relation Between the Generalized Pauli–Villars and the Covariant Regularizations

In this section, some aspects of the comparison that has been established recently by Fujikawa between the generalized Pauli–Villars regularization and the covariant regularization of composite current operators will be investigated, in particular, the question of the choice of regulator, satisfying appropriate conditions [188]. The notion of zeta function of the operators (see Chap. 1) is basic in the discussion. While developing the method, some basic formulas that are useful in physical applications of the theory will be given.

The aim of this section is to show how deeply the zeta function regularization method pervades all the different, alternative regularization procedures that are being used in gauge theory. We see this in the example provided by the interesting

E. Elizalde, *Ten Physical Applications of Spectral Zeta Functions*,
Lecture Notes in Physics 855,
DOI 10.1007/978-3-642-29405-1_7, © Springer-Verlag Berlin Heidelberg 2012

scheme proposed by Frolov and Slavnov of a generalized Pauli–Villars regularization of chiral gauge theory [55] (see also [189, 190]). As clearly pointed out by K. Fujikawa, a formal introduction of an infinite number of regulator fields in the Lagrangian does not specify the method completely and a most fundamental issue in this new regularization is to define how to sum over the contributions coming from the infinite number of regulator fields [191]. By reformulating the generalized Pauli–Villars regularization as a regularization of composite current operators, Fujikawa has proven that an explicit choice for the sum of contributions of the infinite fields results, essentially, in a corresponding specific selection of a regulator, in the language of covariant regularization [192, 193]. (Let us recall that the calculational scheme of covariant anomalies was introduced as an original, conveniently simple method in the path integral formulation of anomalous identities but has not been implemented at the Lagrangian level.)

The chiral gauge theory to regularize is characterized by the Lagrangian

$$\mathcal{L} = \frac{i}{2}\bar{\psi}\slashed{D}(1 + \gamma_5)\psi, \tag{7.1}$$

with

$$\slashed{D} = \gamma^\mu\big(\partial_\mu - igA_\mu(x)\big) = \gamma^\mu\big(\partial_\mu - igA_\mu^a(x)\mathbb{T}^a\big), \tag{7.2}$$

$\mathbb{T}^a$ being the Hermitian generators of a compact semi-simple group: $[\mathbb{T}^a, \mathbb{T}^b] = f^{abc}\mathbb{T}^c$, $\mathrm{tr}(\mathbb{T}^a\mathbb{T}^b) = \frac{1}{2}\delta^{ab}$. The gauge field $A_\mu(x)$ is treated mainly as a background field. The starting point of the generalized Pauli–Villars regularization is to introduce two sets of infinite dimensional mass matrices, M and M', in the following way

$$\mathcal{L} = i\bar{\psi}\slashed{D}\psi - \bar{\psi}_L M\psi_R - \bar{\psi}_R M^\dagger\psi_L + i\bar{\phi}\slashed{D}\phi - \bar{\phi}M'\phi, \tag{7.3}$$

where $\psi_R = \frac{1}{2}(1 + \gamma_5)\psi$, $\psi_L = \frac{1}{2}(1 - \gamma_5)\psi$, and

$$M = \begin{pmatrix} 0 & 2 & 0 & 0 & \cdots \\ 0 & 0 & 4 & 0 & \cdots \\ 0 & 0 & 0 & 6 & \cdots \\ \vdots & \vdots & \vdots & \vdots & \ddots \end{pmatrix}\Lambda, \qquad M' = \begin{pmatrix} 1 & 0 & 0 & 0 & \cdots \\ 0 & 3 & 0 & 0 & \cdots \\ 0 & 0 & 5 & 0 & \cdots \\ \vdots & \vdots & \vdots & \vdots & \ddots \end{pmatrix}\Lambda, \tag{7.4}$$

Λ being a parameter with dimensions of mass. Correspondingly, the fields ψ and ϕ have an infinite number of components, each of them being a conventional anticommuting (resp. commuting) four-component Dirac field. The details of the procedure (with explicit calculations and examples) can be found in the references mentioned above. Let us here only recall the essential steps that have lead to the connection of this regularization with Fujikawa's one. The main issue is to perform the operator expansion

$$\frac{1}{i\slashed{D}} = \frac{1}{i\slashed{\partial} + g\slashed{A}} = \frac{1}{i\slashed{\partial}} + \frac{1}{i\slashed{\partial}}(-g\slashed{A})\frac{1}{i\slashed{\partial}} + \frac{1}{i\slashed{\partial}}(-g\slashed{A})\frac{1}{i\slashed{\partial}}(-g\slashed{A})\frac{1}{i\slashed{\partial}} + \cdots, \tag{7.5}$$

and, after rewriting (7.3) as

$$\mathcal{L} = i\bar{\psi}\mathcal{D}\psi + i\bar{\phi}\mathcal{D}'\phi, \tag{7.6}$$

with

$$\mathcal{D} \equiv \not{D} + iM\frac{1+\gamma_5}{2} + iM^\dagger\frac{1-\gamma_5}{2}, \qquad \mathcal{D}' \equiv \not{D} + iM', \tag{7.7}$$

the additional expansions

$$
\begin{aligned}
\frac{1}{\mathcal{D}} &= \frac{1}{\mathcal{D}^\dagger\mathcal{D}}\mathcal{D}^\dagger = \frac{1}{\not{D}^2 + \frac{1}{2}M^\dagger M(1+\gamma_5) + \frac{1}{2}MM^\dagger(1-\gamma_5)}\mathcal{D}^\dagger \\
&= \left(\frac{1+\gamma_5}{2}\frac{1}{\not{D}^2 + M^\dagger M} + \frac{1-\gamma_5}{2}\frac{1}{\not{D}^2 + MM^\dagger}\right) \\
&\quad \cdot \left(\not{D} - iM^\dagger\frac{1+\gamma_5}{2} - iM\frac{1-\gamma_5}{2}\right), \\
\frac{1}{\mathcal{D}'} &= \frac{1}{(\mathcal{D}')^\dagger\mathcal{D}'}(\mathcal{D}')^\dagger = \frac{1}{\not{D}^2 + (M')^2}(\not{D} - iM').
\end{aligned}
\tag{7.8}
$$

With this, one obtains [191]

$$
\begin{aligned}
&\mathrm{tr}\left[-i\mathbb{T}^a\gamma^\mu\left(\frac{1}{\mathcal{D}} - \frac{1}{\mathcal{D}'}\right)\delta(x-y)\right] \\
&= \frac{1}{2}\mathrm{tr}\left[-i\mathbb{T}^a\gamma^\mu\sum_{n=-\infty}^{\infty}\frac{(-1)^n\not{D}^2}{\not{D}^2 + (n\Lambda)^2}\frac{1}{\not{D}}\delta(x-y)\right] \\
&\quad + \frac{1}{2}\mathrm{tr}\left[-i\mathbb{T}^a\gamma^\mu\gamma_5\frac{1}{\not{D}}\delta(x-y)\right] \\
&= \frac{1}{2}\mathrm{tr}\left[\mathbb{T}^a\gamma^\mu f(\not{D}^2/\Lambda^2)\frac{1}{i\not{D}}\delta(x-y)\right] \\
&\quad + \frac{1}{2}\mathrm{tr}\left[\mathbb{T}^a\gamma^\mu\gamma_5\frac{1}{i\not{D}}\delta(x-y)\right],
\end{aligned}
\tag{7.9}
$$

where the function $f(x^2)$ is defined as the doubly infinite sum

$$f(x^2) \equiv \sum_{n=-\infty}^{+\infty}\frac{(-1)^n x^2}{x^2 + (n\Lambda)^2} = \frac{\pi x/\Lambda}{\sinh(\pi x/\Lambda)}. \tag{7.10}$$

Under the viewpoint of covariant regularization [192, 193], the function $f(x^2)$ is simply to be considered as a regulator that satisfies the conditions

$$
\begin{aligned}
f(0) &= 1, \\
x^2 f'(x^2) &= 0, \quad \text{for } x \to 0, \\
f(+\infty) &= f'(+\infty) = f''(+\infty) = \cdots = 0, \\
x^2 f'(x^2) &\to 0, \quad \text{for } x \to \infty.
\end{aligned}
\tag{7.11}
$$

Thus, following Fujikawa [191], the essence of the generalized Pauli–Villars regularization (7.3) is summarized in terms of the following relations for the regularized currents:

$$
\left\langle \bar{\psi}(x) \mathbb{T}^a \gamma^\mu \frac{1+\gamma_5}{2} \psi(x) \right\rangle_{PV}
$$

$$
= \frac{1}{2} \lim_{y \to x} \left\{ \mathrm{tr}\left[\mathbb{T}^a \gamma^\mu f\left(\slashed{D}^2/\Lambda^2\right) \frac{1}{i\slashed{D}} \delta(x-y) \right] + \mathrm{tr}\left[\mathbb{T}^a \gamma^\mu \gamma_5 \frac{1}{i\slashed{D}} \delta(x-y) \right] \right\},
$$

$$
\left\langle \bar{\psi}(x) \gamma^\mu \frac{1+\gamma_5}{2} \psi(x) \right\rangle_{PV}
$$

$$
= \frac{1}{2} \lim_{y \to x} \left\{ \mathrm{tr}\left[\gamma^\mu f\left(\slashed{D}^2/\Lambda^2\right) \frac{1}{i\slashed{D}} \delta(x-y) \right] + \mathrm{tr}\left[\gamma^\mu \gamma_5 \frac{1}{i\slashed{D}} \delta(x-y) \right] \right\}, \tag{7.12}
$$

$$
\left\langle \bar{\psi}(x) \gamma^\mu \gamma_5 \frac{1+\gamma_5}{2} \psi(x) \right\rangle_{PV}
$$

$$
= \frac{1}{2} \lim_{y \to x} \left\{ \mathrm{tr}\left[\gamma^\mu \gamma_5 f\left(\slashed{D}^2/\Lambda^2\right) \frac{1}{i\slashed{D}} \delta(x-y) \right] + \mathrm{tr}\left[\gamma^\mu \frac{1}{i\slashed{D}} \delta(x-y) \right] \right\},
$$

$$
\left\langle \bar{\psi}(x) \frac{i}{2} \overleftrightarrow{\slashed{D}} \frac{1+\gamma_5}{2} \psi(x) \right\rangle_{PV} = \frac{1}{2} \lim_{y \to x} \mathrm{tr}\left[f\left(\slashed{D}^2/\Lambda^2\right) \delta(x-y) \right].
$$

All the one-loop diagrams are generated from the (partially) regularized currents in (7.12). In other words, (7.11) and (7.12) retain all the information encoded in the generalized Pauli–Villars regularization (7.3). These equations summarize, therefore, the basic results of the new method.

Let us now discuss the basic point in this comparison of the generalized Pauli–Villars and the covariant regularization [188]. If we look at the final expressions (7.12) in the sense that to regularize just means having to choose a specific form for the regulator function $f(x^2)$, with the only requirement that it has to satisfy the conditions (7.11)—what is indeed a quite common point of view—the amount of possibilities that one has in hand for choosing $f(x^2)$ is literally infinite. As has been clearly explained in the first chapter (see also [194])—and has been illustrated through several examples—even if we restrict ourselves just to the class of what can be called 'analytic continuation procedures', this number is still infinite. However (and this is what makes the zeta-function method emerge as unique among all others), if we believe in the *rationale*, i.e. in the mathematical consistency step by step of the above derivation—starting from the generalized Pauli–Villars setup—then we do not have available any more the possibility to choose for the doubly infinite sum (7.10) *any* arbitrary regulator $f(x^2)$ just satisfying the conditions (7.11). This would be equivalent to throwing away all the previous, well constructed derivation and to starting, as an *Ansatz*, from (7.12) with $f(x^2)$ any convenient regulator. (Of course, one can think and proceed in such way, but this is not what the generalized Pauli–Villars regularization is telling us to do.)

And thus we come to the crucial idea: by proceeding consistently one arrives at (7.9), but the double sum here is a perfectly well-defined object and not 'any convenient regulator $f(x^2)$'. In fact, there is a precise, rigorous mathematical theory underlying all the processes of calculation of determinants and traces of differential operators of the type that appear in gauge theories, which starts by introducing the concept of zeta function of the operator (see Chap. 1, last section). Of course, as we have seen, this theory is not without exceptions and some technical problems can appear. But this is not the case in our discussion here: the whole derivations in [191] and [55] satisfy the general premises of the theory, and the expression (7.10) as a sum of an infinite number of terms is the unique result prescribed by this theory, in that particular case. In fact, (7.10) is just a particularly simple example of the general expressions that are obtained by making use of the zeta-function regularization theorem [48, 51, 109] (see Chap. 2). In the case of a series of the type $\sum_{n=0}^{\infty}[a(n+c)^{\alpha}+q]^{-s}$, one proceeds by using the Mellin transform, by expanding the exponent into a power series and, finally, by interchanging the order of summation of the two infinite series, as follows (the systematic procedure can be found in Chap. 2):

$$\sum_{n=0}^{\infty}[a(n+c)^{\alpha}+q]^{-s} = \frac{1}{\Gamma(s)}\sum_{n=0}^{\infty}\int_{0}^{\infty}dt\,t^{s-1}\exp\{-[a(n+c)^{\alpha}+q]t\}$$

$$= \frac{1}{\Gamma(s)}\sum_{n=0}^{\infty}\int_{0}^{\infty}dt\sum_{m=0}^{\infty}\frac{(-1)^{m}}{m!}[a(n+c)^{\alpha}]^{m}t^{m+s-1}e^{-qt}$$

$$\simeq \sum_{m=0}^{\infty}\frac{(-1)^{m}\Gamma(m+s)}{m!\Gamma(s)q^{s}}\left(\frac{a}{q}\right)^{m}\zeta_{H}(-\alpha m,c)+\Delta. \qquad (7.13)$$

In this way one obtains the analytical continuation of the original sum, which is defined only for $\mathrm{Re}\,s > s_0$ (s_0 is the abscissa of convergence), to the whole complex s-plane, as a meromorphic function that is given, in general, under the form of a series (convergent or asymptotic) of ordinary (Riemann or Hurwitz) zeta functions. The non-trivial step in this process is the commutation of the order of the summations, that has originated multiple errors in the physical literature [10, 11, 14]. Such commutations involve non-simple contour integrations on the complex plane, as shown before. This is how additional contributions appear, denoted in (7.13) by Δ (the whole procedure is extensively described in Chaps. 2 and 3 of this book).

To summarize, if one adheres to this theory then there is no possible second interpretation of the doubly infinite sum (7.10). In fact, the expression for $f(x^2)$ is just a particular case of the formulas above. To wit

$$
f(x^2) = \sum_{n=-\infty}^{+\infty} \frac{(-1)^n x^2}{x^2 + \Lambda^2 n^2}
$$

$$
= x^2 \left[\sum_{n=2k} (x^2 + \Lambda^2 n^2)^{-1} - \sum_{n=2k+1} (x^2 + \Lambda^2 n^2)^{-1} \right]
$$

$$
= x^2 \left[\frac{1}{2} g \left(\frac{x}{2} \right) - g(x) \right], \tag{7.14}
$$

with

$$
g(x) \equiv \sum_{n=-\infty}^{+\infty} (x^2 + \Lambda^2 n^2)^{-1} = \frac{\pi}{\Lambda x} + \frac{4\pi}{\Lambda \sqrt{\Lambda x}} \sum_{n=1}^{\infty} \sqrt{n}\, K_{1/2}(2\pi n x/\Lambda), \tag{7.15}
$$

which is the particular case $c = s = 1$, $a = \Lambda^2$, $q = x^2$, of (2.93), so that

$$
f(x^2) = \pi \left(\frac{2x}{\Lambda} \right)^{3/2} \sum_{n=1}^{\infty} \sqrt{n} \left[K_{1/2}(\pi n x/\Lambda) - \sqrt{2} K_{1/2}(2\pi n x/\Lambda) \right]. \tag{7.16}
$$

And from $K_{1/2}(z) = \sqrt{\frac{\pi}{2z}} e^{-z}$ we get, finally,

$$
f(x^2) = \frac{2\pi x}{\Lambda} \sum_{n=1}^{\infty} \left(\frac{e^{-\pi x/\Lambda}}{1 - e^{-\pi x/\Lambda}} - \frac{e^{-2\pi x/\Lambda}}{1 - e^{-2\pi x/\Lambda}} \right) = \frac{\pi x/\Lambda}{\sinh(\pi x/\Lambda)}. \tag{7.17}
$$

This calculation is certainly simple. It can be extended to all similar expression that appear in this regularization, where x stands for an operator, what means that analytical continuation must be performed in a nested way (but this is just perfectly correct within the method). According to the previous discussion the exercise above has to be considered as just the visible top of a rather huge iceberg, e.g., as a very particular case of the powerful procedure which consists in defining the traces and determinants of the differential operators that appear in gauge theories in a rigorous way, by means of the introduction of the corresponding zeta function (Chap. 1).

7.2 The Casimir Energy Corresponding to a Piecewise Uniform String

In this section, the Casimir energy for the transverse oscillations of a piecewise uniform closed string is calculated. The string consists of two parts (that we will call

I and II), each having in general different tension and mass densities, but adjusted in such a way that the velocity of sound always equals the velocity of light. This model was introduced by I. Brevik and H.B. Nielsen, and here we will describe new developments of the theory, in particular, a quite simple regularization of the energy density. Using the technique introduced by N.G. van Kampen, B.R.A. Nijboer, and K. Schram, the Casimir energy is written as a contour integral, from which the energy can be readily calculated, for arbitrary length $s = L_{II}/L_I$ and tension $x = T_I/T_{II}$ ratios. Also, the finite temperature version of the theory will be constructed (see also [195, 196]).

Consider, in Minkowski space, a closed string of length L composed of two parts, of lengths L_I and L_{II}, respectively [36]. The tensions, T_I and T_{II}, and mass densities, ρ_I and ρ_{II}, corresponding to the two pieces are in general different, but they will be required to satisfy the condition that the sound velocity be always equal to the light velocity, i.e.

$$v_s = (T_I/\rho_I)^{1/2} = (T_{II}/\rho_{II})^{1/2} = c. \tag{7.18}$$

The purpose here is to study the Casimir energy associated with the transverse oscillations of this piecewise uniform string. It turns out that this model is, at least from a formal point of view, quite an interesting one. Some calculations can be carried out without encountering the annoying divergences in the regularized result, which so often plague Casimir calculations when the geometry is nontrivial (curved boundary surfaces, typically). A basic point in this context is condition (7.18), which renders the string relativistically invariant, and is analogous to requiring the refractive index $(\epsilon\mu)^{1/2}$ to be equal to unity. In this sense, it is basically of the same kind as the color medium proposed by T.D. Lee [197] for the region exterior to a hadron. If, by contrast, condition (7.18) were abandoned, then divergence problems would certainly show up in the formalism.

From a physical point of view, there is founded hope that this simple model can help us to understand the issue of the energy of the vacuum state in two-dimensional quantum field theories in general, a quite compelling goal. The model was introduced by Brevik and Nielsen in an earlier paper [36] (see also the recent paper [196]), to which the reader is referred for specific details [195]. There, the zero-point energy was regularized by means of an exponential cutoff. It was also pointed out, that the use of more formal regularization procedures—such as the ζ-function method—might lead to delicate problems; in particular, that (a naively-minded, straightforward) use of the Riemann ζ-function could lead to an incorrect result. This problem was reconsidered and solved in an elegant way by Li, Shi, and Zhang [37]. The final results obtained in [37] were in agreement with those of [36]. The whole situation concerning the use of the zeta function regularization procedure in this and similar cases is discussed in [194] in great detail.

The main results that will be described in this section are the following.

1. The Casimir energy at zero temperature, $T = 0$, will be found as a double function of the length ratio $s \equiv L_{II}/L_I$, for any value of s, and of the tension ratio

$x \equiv T_I / T_{II}$, for arbitrary x. To compare, in Ref. [36] the solution was given explicitly for a few lowest integer values of s and a few selected values of x, only (and this after considerable work). To achieve our goal, we shall employ here a quite elegant technique, based on a well known theorem of complex analysis, and which was first introduced—in a context related to the present one—by van Kampen, Nijboer, and Schram some years ago [198]. It consists in rewriting the Casimir energy under the form of a very simple contour integral. This technique, when applied to the present problem must be used with some care, in order to avoid a nonphysical divergence in the form of a surface term. However, as we shall see below, the suppression of the surface term can be done consistently, through a proper choice of the dispersion function for the system.

2. When using this technique, it becomes unnecessary to take the degeneracies of the eigenfrequencies of the system into account explicitly. This comes as a useful bonus. The reason is that the degeneracies precisely correspond to the multiplicities of the zeros which appear in the *argument principle* (cf. (7.23) below). This fact makes the final theory much more simple, as compared with the original procedure of finding and counting the roots of the dispersion equation, that had been used in Ref. [36] (see also [37]).

3. The Casimir energy for the string is calculated for finite temperature, $T \neq 0$, also, and the analytic approximation for high T is worked out. This high temperature limit provides an immediate check of the procedure, since it is easy to find analytically.

7.2.1 The Zero Temperature Theory

We shall use the same notation as in Ref. [36]. The total length of the string is $L = L_I + L_{II}$. Denoting by x the ratio between the two tensions,

$$x = \frac{T_I}{T_{II}}, \tag{7.19}$$

the dispersion equation can be written as

$$\frac{4x}{(1-x)^2} \sin^2 \left(\frac{\omega L}{2c} \right) + \sin \left(\frac{\omega L_I}{c} \right) \sin \left(\frac{\omega L_{II}}{c} \right) = 0. \tag{7.20}$$

The physical requirements behind this equation are (i) continuity of the transverse displacements across the two junctions, and (ii) continuity of the transverse elastic forces across the junctions. Since (7.20) is invariant under the substitution $x \to 1/x$, we can simply take $0 \leq x \leq 1$ in what follows (the case $x = 0$ must be considered with some care).

The Casimir energy of the system, E, is constructed such that it describes the nonhomogeneity of the string only, and is thus required to vanish for a uniform

Fig. 11 Integration contour
in the complex ω plane

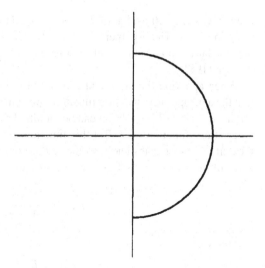

string. Therefore, E is equal to the zero-point energy E_{I+II} for the two parts, minus the zero-point energy for the uniform string, i.e.

$$E = E_{I+II} - E_{uniform}. \tag{7.21}$$

It should be noted that, when subtracting off $E_{uniform}$, it is completely irrelevant whether the uniform string is composed of type I or type II material. The physical reason for this is that the frequency spectrum for the uniform string is independent of the type of material, as long as the velocity of sound is the same, and thus it is in the present case a consequence of (7.18).

The zero-point energy of the composite string is

$$E_{I+II} = \frac{\hbar}{2} \sum \omega_n, \tag{7.22}$$

where the sum goes over all eigenstates, with account to their degeneracy. Stationarity of the oscillating system implies that all the eigenfrequencies ω_n have to be real. We can let all the ω_n be positive, left-moving waves being associated with negative *wave numbers* (not frequencies). And here comes the first important idea: the sum (7.22) can be written in the form of a contour integral by means of a well known mathematical theorem called the argument principle [131, 199]. It states that any meromorphic function satisfies the equation

$$\frac{1}{2\pi i} \oint \omega \frac{d}{d\omega} \ln g(\omega) \, d\omega = \sum \omega_0 - \sum \omega_\infty, \tag{7.23}$$

where ω_0 denotes the zeros and ω_∞ the poles of $g(\omega)$ lying inside the integration contour as shown in Fig. 11. The argument principle, a form of Cauchy's theorem. In the end, the radius R of the contour in (7.23) is allowed to go to infinity (as is usually the case for theorems of this kind on the complex plane). The multiplicity of the zeros, as well as that of the poles, are automatically taken care of by the two sums in (7.23). The argument principle was first applied to the Casimir theory (in the standard configuration with two parallel plates) by van Kampen, Nijboer, and Schram [198].

When applying the argument principle to our present problem, we first notice that the appropriate dispersion function $g(\omega)$ must essentially be the function on the left hand side of (7.20)—but it can be modified by a factor not depending on ω, e.g., an arbitrary function of x. But this function $g(\omega)$ has no poles, therefore, the last term in (7.23) vanishes and thus the crosses on the real axis in Fig. 11 refer to the zeros of $g(\omega)$ only. As in Ref. [36] we introduce the function

$$F(x) = \frac{4x}{(1-x)^2}, \tag{7.24}$$

and the variable

$$s = \frac{L_{II}}{L_I}, \tag{7.25}$$

to denote the ratio of the lengths of the two pieces of the string. For definiteness we shall take L_I to be the smaller of the two pieces, so that $s \geq 1$. For the dispersion function of the composite system, we now make the *Ansatz*:

$$g(\omega) = \frac{F(x)\sin^2[(s+1)\omega L_I/(2c)] + \sin(\omega L_I/c)\sin(s\omega L_I/c)}{F(x)+1} \tag{7.26}$$

(see the remarks following (7.29) and (7.32) below). For given values of x and of the total length L, this expression is invariant under the substitution $s \to 1/s$. Thus, the above restriction to values of $s \geq 1$ represents no loss in generality. By making use of the argument principle, for the composite system we have the zero point energy

$$E_{I+II} = \frac{\hbar}{4\pi i} \oint \omega \frac{d}{d\omega} \ln|g(\omega)| \, d\omega, \tag{7.27}$$

with the function $g(\omega)$ given by (7.26). In writing this expression, we have taken advantage of the following important correspondence between degeneracy of the eigenfrequencies ω_n and multiplicity of the zeros of $g(\omega)$. As noted in connection with (7.22), it is in general necessary to take into account degeneracies when summing over all states. In Ref. [36] the degeneracies were actually counted explicitly, for each branch of the dispersion equation, in the cases of low integer s for which the solution was worked out. Within the present approach, the handling of the degeneracy problem is, however, much more easy since the argument principle (7.23), as we have seen, already takes into account the multiplicity of the zeros. There exists a one to one correspondence between the degeneracy of the eigenfrequencies and the multiplicity of the zeros. Therefore, the degeneracies are built in automatically, in the integral (7.27).

7.2.2 Regularized Casimir Energy and Numerical Results

In spite of these interesting properties, one should notice that, as it stands, (7.27) is not a useful expression. In fact, it is not difficult to see that the contribution of the curved part, the contour of radius R (Fig. 11), to the integral (7.27) grows without bound as $R \to \infty$. Since, in general, in order to take into account all the modes in the series (7.22), we must send R to infinity, it follows that a divergence is hidden in the curved contour at infinity. What one has to do is to subtract off the energy of the uniform string. This corresponds to $x = 1$ (the value of s need not be specified). Since $F(x) \to \infty$ as $x \to 1$, we obtain using (7.26)

$$E_{uniform} = \frac{\hbar}{4\pi i} \oint \omega \frac{d}{d\omega} \ln \sin^2 \left[\frac{(s+1)\omega L_I}{2c} \right] d\omega, \tag{7.28}$$

and thus the Casimir energy follows from (7.21)

$$E = \frac{\hbar}{4\pi i} \oint \omega \frac{d}{d\omega} \ln \left| \frac{F(x) + \frac{\sin(\omega L_I/c)\sin(s\omega L_I/c)}{\sin^2[(s+1)\omega L_I/(2c)]}}{F(x) + 1} \right| d\omega. \tag{7.29}$$

It is easy to see that when the two pieces have the same length, $L_I = L_{II}$ (i.e., $s = 1$), then $E = 0$, irrespective of the value of x. This is just as it should—according to the detailed considerations in Ref. [36]—a fact that actually was the reason behind our particular choice (7.26) for the dispersion function. Notice also, as a corollary, that (7.29) yields $E = 0$ when $x = 1$.

The contribution from the semicircle of Fig. 11 to the integral in (7.29) is now seen to vanish in the limit $R \to \infty$, and the remaining integral along the imaginary axis ($\omega = i\xi$) is integrated by parts, while keeping R finite and taking advantage of the symmetry of the integrand about the origin. We get

$$E = -\frac{\hbar}{2\pi} R \ln \left| \frac{F(x) + \frac{\sinh(RL_I/c)\sinh(sRL_I/c)}{\sinh^2[(s+1)RL_I/(2c)]}}{F(x) + 1} \right|$$

$$+ \frac{\hbar}{2\pi} \int_0^R \ln \left| \frac{F(x) + \frac{\sinh(\xi L_I/c)\sinh(s\xi L_I/c)}{\sinh^2[(s+1)\xi L_I/(2c)]}}{F(x) + 1} \right| d\xi. \tag{7.30}$$

Here, the boundary term is seen to vanish when $R \to \infty$, and thus we obtain, finally,

$$E = \frac{\hbar}{2\pi} \int_0^\infty \ln \left| \frac{F(x) + \frac{\sinh(\xi L_I/c)\sinh(s\xi L_I/c)}{\sinh^2[(s+1)\xi L_I/(2c)]}}{F(x) + 1} \right| d\xi. \tag{7.31}$$

We assume that the total length, L, and the tension ratio, x, are given quantities. Therefore, $F(x)$ is known, and (7.31) gives E as a function of the length ratio s. However, this expression is easily calculable on a computer, and we can equally well give E as a double function of x and s (Fig. 12). As a corollary, we have

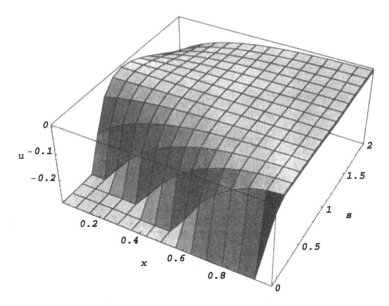

Fig. 12 Three-dimensional plot of the Casimir energy as obtained from any of the two above expressions for the energy, since both figures are visually undistinguishable. The magnitude $u \equiv EL/(\hbar c)$ is plotted versus the tension ratio $x \in [0, 1]$ (*first axis*) and length ratio $s \in [0, 2]$ (*second axis*). The Casimir energy is generically seen to be negative. Only for equal lengths ($s = 1$) is the maximum energy $E = 0$ attained

checked that, in the special case $x \rightarrow 0$, (7.31) gives results which are in agreement with the analytic (well known) expression

$$E = -\frac{\pi \hbar c}{24L}\left(s + \frac{1}{s} - 2\right) \qquad (7.32)$$

derived in Ref. [36].

A remark is in order, concerning our inclusion of the factor $F(x)[F(x) + 1]^{-1}$ in (7.26). Had we *not* introduced this factor in $g(\omega)$, namely had we taken

$$g(\omega) = \sin^2\left[\frac{(s + 1)\omega L_I}{2c}\right] + F(x)^{-1} \sin\left(\frac{\omega L_I}{c}\right) \sin\left(\frac{s\omega L_I}{c}\right), \qquad (7.33)$$

we would have obtained

$$E = -\frac{\hbar}{2\pi} \int_0^\infty \xi \frac{d}{d\xi} \ln\left|1 + \frac{\sinh(\xi L_I/c) \sinh(s\xi L_I/c)}{F(x) \sinh^2[(s + 1)\xi L_I/(2c)]}\right| d\xi, \qquad (7.34)$$

which means that, for $x = 0$,

$$E(x = 0) = -\frac{\hbar c}{2\pi L_I} \int_0^\infty t \frac{d}{dt} \ln\left|\frac{\sinh(t) \sinh(st)}{\sinh^2[(s + 1)t/2]}\right| dt. \qquad (7.35)$$

Thus, (7.34) is valid for the whole range of possible values of x, $0 \le x \le 1$.

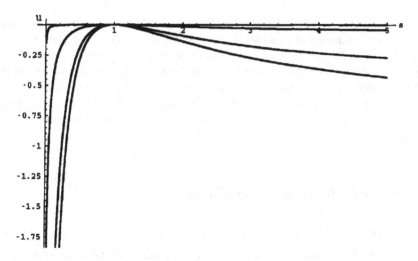

Fig. 13 Several x-sections of the previous plot are depicted, to show how the energy varies as a function of the length ratio s, for several fixed values of x, namely $x = 0$, $x = 0.1$, $x = 0.5$, and $x = 0.9$, respectively. All curves give the magnitude $u \equiv EL/(\hbar c)$ as a function of s in the range $s \in [0, 5]$

As an additional numerical test, by introducing an upper cutoff, K,

$$E(K) \equiv -\frac{\hbar c}{2\pi L_I} \int_0^K t \frac{d}{dt} \ln \left| 1 + \frac{\sinh(t)\sinh(st)}{F(x)\sinh^2[(s+1)t/2]} \right| dt, \qquad (7.36)$$

it turns out that for values of K between say $K \simeq 30$ and $K \simeq 10^4$, the result for $E(K)$ does not change numerically (to 10 digits) and coincides with the value given by (7.31). This has been checked for the whole range of values of x, $0 \le x \le 1$, and s, $s \ge 1$. The problem arises if one performs a partial integration in (7.34): then the resulting first term (i.e., the surface term) turns out to be divergent. In other words, the cutoff K must be kept. This is the drawback associated with expression (7.34).

As regards the symmetric behavior with respect to the values of s, i.e., the coincidence of the results corresponding to s and $1/s$, for any s, it is easily seen to hold for (7.34) explicitly (as for (7.31)). Therefore, our choice above of letting s to be restricted to values ≥ 1, does not represent any loss in generality. These conclusions are very nice indeed, and give sense to the introduction of the x-dependent factor in $g(\omega)$, and to the final formula (7.31) itself as a very simple, regularized expression for the Casimir energy. The numerical results are collected in Figs. 12 and 13. Figure 12 is a three-dimensional plot of the Casimir energy as obtained from (7.36) or (7.31) (both figures are visually undistinguishable). Specifically, it shows how the magnitude $EL/(\hbar c)$ varies versus $x \in [0, 1]$ (first axis) and $s \in [0, 2]$ (second axis). The Casimir energy is generically seen to be negative. Only for equal lengths ($s = 1$) is the maximum energy $E = 0$ found, irrespective of the value of x. In Fig. 13, several x-sections of the previous plot are depicted. It is shown here how the energy varies as a function of s, for several fixed values of x, corresponding to

$x = 0$, $x = 0.1$, $x = 0.5$, and $x = 0.9$, respectively. All of them give the magnitude $EL/(\hbar c)$ as a function of s for the range $s \in [0, 5]$. These figures must be compared with the one of Ref. [36] where, as mentioned above, the solution (obtained after laborious numerical calculation) was given explicitly for a few lowest integer values of s and a few selected values of x. For low integers s up to $s = 7$ and corresponding values for x, we have checked that the results calculated from (7.31) are in agreement with those of Ref. [36]. The advantages of the present procedure are however unquestionable.

7.2.3 The Finite Temperature Theory

1. *General Formalism*

Once the $T = 0$ theory is established, we can readily generalize the situation to the case of finite temperatures, by means of the substitution [200]

$$\hbar \int_0^\infty d\xi \longrightarrow 2\pi k_B T \sum_{n=0}^{\infty}{}', \tag{7.37}$$

where the prime means that the contribution for $n = 0$ has to be taken with half weight. The discrete Matsubara frequencies are $\xi_n = 2\pi n k_B T/\hbar$, $n = 0, 1, 2, \ldots$. From (7.31) we then get for the Casimir energy at an arbitrary temperature, T,

$$E(T) = k_B T \sum_{n=0}^{\infty}{}' \ln \left| \frac{F(x) + \frac{\sinh(\xi_n L_I/c)\sinh(s\xi_n L_I/c)}{\sinh^2[(s+1)\xi_n L_I/(2c)]}}{F(x) + 1} \right|. \tag{7.38}$$

If the string is uniform, $x = 1$ or $F(x) \to \infty$, then (7.38) yields $E(T) = 0$. This is just as we would expect, since the Casimir energy is intended to describe the effect of the inhomogeneity of the string only. Moreover, also the case $L_I = L_{II}$ is seen to yield $E(T) = 0$, irrespective of the value of x. Both these properties, noted earlier for the $T = 0$ theory, do therefore carry over to the case of arbitrary T.

There are two characteristic frequencies in our system:

1. The thermal frequency, ω_T, which can be defined by $\hbar \omega_T = k_B T$. We may observe that ω_T is related to the $n = 1$ Matsubara frequency ξ_1 through $\omega_T = \xi_1/(2\pi)$.
2. The geometric frequency ω_{geom}, associated with the geometry of the string. We may choose to define ω_{geom} in terms of L_I as fundamental length: $\omega_{geom} = 2\pi c/L_I$.

There would also be a third characteristic frequency in the problem, if the *microstructure* of the string were to be taken into account. It would correspond to the absorption frequency(ies) in the dispersion equation for an ordinary dielectric material. But we shall leave out of consideration microstructure effects here. The limiting cases of high and low temperatures are conveniently discussed in terms of the ratio between ω_T and ω_{geom}.

2. *High Temperatures*

Assume that

$$\frac{\omega_T}{\omega_{geom}} \geq 1. \tag{7.39}$$

This is the natural condition for applying the high-temperature approximation. Generally, high temperatures are associated with contributions coming from low Matsubara frequencies only. In our case, we have

$$\frac{\xi_n L_I}{c} = 4\pi^2 n \frac{\omega_T}{\omega_{geom}} \gg 1, \tag{7.40}$$

even for the lowest non-vanishing frequency ($n = 1$), so that $\sinh(\xi_n L_I/c) \simeq (1/2)\exp(\xi_n L_I/c)$, etc., in (7.38). The contribution to $E(T)$ from $n \geq 1$, in this approximation, is accordingly seen to vanish and we are left with just the $n = 0$ term. The result is

$$E(T) = \frac{k_B T}{2} \ln \left| \frac{F(x) + 4s/(s+1)^2}{F(x) + 1} \right|. \tag{7.41}$$

The main corrections to this expression come of course from $n = 1$, and are of order $k_B T \exp(-2\xi_1 L_I/c)$. (7.41) is seen to be a classical result, since it is independent of $\hbar$. Notice that $E(T) \leq 0$ always, the equality sign being valid when $s = 1$, as we have pointed out above.

Similarly to the considerations done in Ref. [36], we can in pictorial terms associate part I of the string with our universe, and part II of it with a mirror universe. If our universe is small and the mirror universe large we get, since $s \to \infty$, the simple expression

$$E(T) = -\frac{k_B T}{2} \ln \left| 1 + F(x)^{-1} \right|. \tag{7.42}$$

To get a feeling of the numerical magnitudes involved here, let us first choose $L_I = 1$ μm, in which case $\omega_{geom} = 2\pi c/L_I = 1.88 \cdot 10^{15}$ s^{-1}. The ratio between the frequencies ω_T and ω_{geom} becomes then

$$\frac{\omega_T}{\omega_{geom}} = 0.70 \cdot 10^{-4} T, \tag{7.43}$$

showing that $T \geq 10^4$ K (i.e., 0.86 eV), if the high temperature approximation is to hold. As another example, let us take the extreme case of $L_I = 1.62 \cdot 10^{-33}$ cm, the Planck length. Then $\omega_{geom} = 1.16 \cdot 10^{44}$ s^{-1}, and

$$\frac{\omega_T}{\omega_{geom}} = 1.13 \cdot 10^{-33} T. \tag{7.44}$$

This case thus requires very high temperatures, $T \geq 10^{33}$ K (i.e., $0.86 \cdot 10^{20}$ GeV), for the high-temperature approximation to be valid.

3. *Low Temperatures*

This limiting case is characterized by

$$\frac{\omega_T}{\omega_{geom}} \ll 1, \tag{7.45}$$

and a large number of Matsubara frequencies comes into play in (7.38). This equation, as it stands, is not written in a convenient form for performing analytical approximations when the temperatures are low. Rather often, when the mathematics is manageable, it is quite useful to exploit the Poisson summation formula, whereby the series over n can be handled approximately without much trouble. However, in the present case the logarithmic summand in (7.38) is too complicated to permit an efficient use of the Poisson formula. At least for practical purposes, the best way to proceed is to deal directly with the series by means of a computer program, even in the case of low frequencies. The necessary number of terms of the series can be added up in a few seconds, to attain any desired precision. A convenient way of writing the series for low T, for making use of these techniques, is the following:

$$E(T) = k_B T \left\{ \frac{1}{2} \ln \left| \frac{F(x) + 4s/(s+1)^2}{F(x) + 1} \right| \right.$$

$$\left. + \sum_{n=1}^{n_T} \ln \left| \frac{F(x) + (1 - e^{-2nb})(1 - e^{-2snb})(1 - e^{-(s+1)nb})^{-2}}{F(x) + 1} \right| \right\}, \tag{7.46}$$

where we have called

$$b = \frac{2\pi k_B T L}{\hbar(s+1)c}, \tag{7.47}$$

and where, for a given temperature T, the sum can be safely truncated at a value n_T such that e.g. $n_T b \simeq 10$.

Table 7 shows, as an example, how the Casimir energy changes with the temperature in the case when $L_I = 1$ μm, $s = 2$, and $F = 1$ (i.e., $x = 0.1716$, according to (7.24)). To distinguish between the low and the high temperature regions, the corresponding values of ω_T/ω_{geom} are given. For low but finite temperatures, the values of the upper limit, n_T, occurring in (7.46), chosen as $n_T = 10/b$, are also given. From the present data, it is clearly seen that the Casimir energy becomes more and more negative as the temperature increases.

We have studied in this section different issues related with a most convenient definition of the Casimir energy for the transverse oscillations of a piecewise uniform string. We have proven that, in fact, the calculations can be carried out in a remarkably easy way. Not only because annoying divergences can be completely avoided in the regularized result, but also because the expressions leading to this

Table 7 Values of the Casimir energy E for some different values of T, assuming $L_I = 1$ μm, $s = 2$, and $F = 1$

	$T(K)$		
	0	10	300
ω_T/ω_{geom}	0	$6.95 \cdot 10^{-4}$	$2.09 \cdot 10^{-2}$
n_T	–	365	13
$E(erg)$	$-3.3770 \cdot 10^{-15}$	$-3.3847 \cdot 10^{-15}$	$-3.3848 \cdot 10^{-15}$

	$T(K)$		
	$3 \cdot 10^3$	$3 \cdot 10^4$	10^6
ω_T/ω_{geom}	0.209	2.09	69.5
n_T	2	–	–
$E(erg)$	$-1.18 \cdot 10^{-14}$	$-1.18 \cdot 10^{-13}$	$-3.94 \cdot 10^{-12}$

finite result are quite simple (see (7.31)), and allow us to calculate the most general case (see Fig. 12). Hence, simplicity is one of the main virtues of the model. Generality of the procedure is another.

Notwithstanding the fact that it is so simple, we do sustain the hope that such a model can actually help us to understand the issue of the energy of the vacuum state in two-dimensional quantum field theories, what is quite a compelling goal in itself. A specific result arising from the above calculations is the Casimir energy at zero temperature, as an explicit double function of the length ratio s and of the tension ratio x, for arbitrary s and x. To this end we have devised an elegant technique, based on the argument theorem of complex analysis. This has led us to a formula which, when applied to the present problem, leads to a final result free of any nonphysical divergences (in particular, of the surface divergence that was to be expected). We have shown in detail, that the suppression of this divergence can be done consistently, by a proper choice of the dispersion function.

Also, it has become unnecessary to take the degeneracies of the eigenfrequencies of the system into account explicitly, because the degeneracies precisely correspond to the multiplicities of the zeros which appear in the argument principle (see (7.23)). This principle, which is a form of the fundamental Cauchy theorem, is also the key point in the new applications of zeta regularization to the calculation of heat-kernel coefficients and determinants of the Laplacian with different boundary conditions that have been mentioned in Chap. 1 (last section, see [56, 64]). The property mentioned above renders the final theory much more simple, as compared with the original procedure of finding and counting the roots of the dispersion equation, that had been used in Ref. [36]. The Casimir energy for the string has been calculated here at finite temperature $T \neq 0$. The analytic approximation for high T has been obtained, and a formula well suited for numerical calculations in the low T limit has been given too. The specific meaning of these limits in terms of the characteristic frequencies of the system has been discussed numerically.

7.3 Zeta and Hadamard Regularizations

There has been a lot of discussion in recent years on the issue of imposing boundary conditions on quantum fields and its relation with the appearance of infinities and the subsequent (physical vs. mathematical) regularization plus renormalization process. This made me quote a famous sentence by Einstein: "*As far as the laws of mathematics refer to reality, they are not certain; and as far as they are certain, they do not refer to reality*", in one of my papers on this issue [201]. Indeed, an interesting example of the deep interrelation between Physics and Mathematics is obtained when trying to impose mathematical boundary conditions on physical quantum fields. This procedure was recently re-examined with care. Comments on that and previous analysis are here provided, together with considerations on the results of the purely mathematical zeta-function method, in an attempt at clarifying this point. Hadamard regularization will be invoked in order to fill the gap between the infinities appearing in the QFT renormalized results and the finite values obtained in the literature with other procedures.

The question, phrased by Eugene Wigner as that of "*the unreasonable effectiveness of mathematics in the natural sciences*" [202] is an old and intriguing one. It goes back to Pythagoras and his school ("*all things are numbers*"), even probably to the Sumerians, and maybe to more ancient cultures, which left no trace. I. Kant and A. Einstein contributed also to this idea with profound reflections, and *mathematical simplicity, and beauty*, have remained for many years crucial ingredients when having to choose among different plausible possibilities. An example of *unreasonable effectiveness* is provided by the regularization procedures in quantum field theory (QFT) based upon analytic continuation in the complex plane (dimensional, heat-kernel, zeta-function regularization, and the like). That one obtains a physical, experimentally measurable, and extremely precise result after these weird mathematical manipulations is, if not unreasonable, certainly very mysterious. For more one highly honorable physicist those remained always illegal practices. Such methods are now full justified and blessed with Nobel Prizes, but more because of the many and very precise experimental checkouts (the *effectiveness*) than for their intrinsic *reasonableness*.

A simple example may be clarifying. Consider the calculation of the zero point energy (vacuum to vacuum transition, the Casimir energy discussed above [203]) corresponding to a quantum operator, H, with eigenvalues λ_n: $E_0 = \langle 0|H|0\rangle = \frac{1}{2}\sum_n \lambda_n$, where the sum over n may involve several continuum and discrete indices. Only in special cases will this sum be convergent. Generically one has a divergent series, to be regularized by different means. The zeta-function method [10, 11, 13, 14]—which stands on solid and flourishing mathematical grounds [204–212]—will interpret it as the value of the zeta function of H: $\zeta_H(s) = \sum_n \lambda_n^{-s}$, at $s = -1$ (we set $\hbar = c = 1$). Generically $\zeta_H(s)$ is only defined as an absolutely convergent series for $\operatorname{Re} s > a_0$ (a_0 an abscissa of convergence), but it can be continued to the whole complex plane, with the possible appearance of poles as only singularities. If $\zeta_H(s)$ has no pole at $s = -1$ then we are done; if it hits a pole, further elaboration is necessary. That the mathematical result one thus gets coincides with the experimental one, constitutes here our specific example of *unreasonable effectiveness of mathematics*.

In fact things do not turn out to be so simple. In an isolated vacuum one cannot assign physical meaning to the *absolute* value of the zero-point energy (we can safely fix it to be zero, by using e.g. the normal ordering prescription). The most simple physical effect must necessarily be an energy difference between two situations, such as a quantum field in curved space as compared with the same field in flat space, or one satisfying BCs on some surface as compared with the same in its absence, etc. This difference, the Casimir energy, is a genuine physical manifestation of the vacuum energy: $E_C = E_0^{BC} - E_0 = \frac{1}{2}(\text{tr } H^{BC} - \text{tr } H)$. This seems at first sight to be clean, unproblematic, at least at the level of the mathematical formulation of the issue.

But at the physical level problems appear. Imposing mathematical boundary conditions (BCs) on physical quantum fields turns out to be a highly non-trivial act. This was discussed in much detail in a paper by Deutsch and Candelas over thirty years ago [213]. These authors quantized electromagnetic and scalar fields in the region near an arbitrary smooth boundary, and calculated the renormalized vacuum expectation value of the stress-energy tensor, to find that the energy density diverges as the boundary is approached. Therefore, regularization and renormalization did not seem to cure the problem with infinities in this case and an infinite *physical* energy was obtained if the mathematical BCs were to be fulfilled. However, the authors argued that surfaces have non-zero depth, and its value could be taken as a handy (dimensional) cutoff in order to regularize the infinities. This approach will be recovered later in this paper. Just two years after Deutsch and Candelas' work, Kurt Symanzik carried out a rigorous analysis of QFT in the presence of boundaries [214]. Prescribing the value of the quantum field on a boundary means using the Schrödinger representation, and Symanzik was able to show rigorously that such representation exists to all orders in the perturbative expansion. He showed also that the field operator being diagonalized in a smooth hypersurface differs from the usual renormalized one by a factor that diverges logarithmically when the distance to the hypersurface goes to zero. This requires a precise limiting procedure and point splitting to be applied. In any case, the issue was proven to be perfectly meaningful within the domains of renormalized QFT. In this case the BCs and the hypersurfaces themselves were treated at a pure mathematical level (zero depth) by using delta functions.

A new approach to the problem has been postulated by Jaffe and collaborators [215–218]. BCs on a field, ϕ, are enforced on a surface, S, by introducing a scalar potential, σ, of Gaussian shape living on and near the surface. When the Gaussian becomes a delta function, the BCs (Dirichlet here) are enforced: the delta-shaped potential kills *all* the modes of ϕ at the surface. For the rest, the quantum system undergoes a full-fledged QFT renormalization, as in the case of Symanzik's approach. The results obtained confirm those of [213] in the several models studied, but the authors report that they do not seem to agree with those of [214, 219–221].

7.3.1 A Zeta-Function Approach

Too often has it been argued that sophisticated regularization methods, as the zeta-function procedure, get rid of infinities in an obscure way (e.g. through analytic continuation), so that, contrary to what happens with cut-offs, one cannot keep trace of the infinities, which are cleared up without control, leading sometimes to erroneous results. One cannot refute a statement of this kind rigorously, but it should be noted that more once (if not always) the discrepancies between the result obtained by using the zeta procedure and other—say cut-off like—approaches have been proven to emerge from a *misuse* of zeta regularization, and *not* to stem from the method itself. When employed properly, the correct results have been recovered (for a number of examples, see [10, 13, 14, 204–212, 222–226]).

Take the most simple case of a scalar field in one dimension, $\phi(x)$, with a BC of Dirichlet type imposed at a point, e.g. $\phi(0) = 0$. We would like to calculate the Casimir energy for this configuration, that is, the difference between the zero point energy corresponding to this field when the BC is enforced, and the zero point energy in the absence of any BC. Taken at face value, both energies are infinite. The regularized difference may still be infinite when the BC point is approached (this is the result in [215–218]) or might turn out to be finite (even zero, which is the result given in some standard books on the subject).

Let us try to understand this discrepancy. We have to add up all energy modes (trace of H). For the mode with energy ω, the field equation reduces to:

$$-\phi''(x) + m^2\phi(x) = \omega^2\phi(x). \tag{7.48}$$

In the absence of a BC, the solutions to the field equation can be labeled by $k = +\sqrt{\omega^2 - m^2} > 0$, as $\phi_k(x) = Ae^{ikx} + Be^{-ikx}$, with A, B arbitrary complex (for the general complex), or as $\phi_k(x) = a\sin(kx) + b\cos(kx)$, with a, b arbitrary real (for the general real solution). Now, when the mathematical BC of Dirichlet type, $\phi(0) = 0$, is imposed, this does not influence at all the eigenvalues, k, which remain exactly the same (as stressed in the literature). However, the *number* of solutions corresponding to each eigenvalue is reduced by one half to: $\phi_k^{(D)}(x) = A(e^{ikx} - e^{-ikx})$, with A arbitrary complex (complex solution), and $\phi_k^{(D)}(x) = a\sin(kx)$, with a arbitrary real (real solution). In other words, the energy spectrum (for omega) that we obtain in both cases is the same, a continuous spectrum $\omega = \sqrt{m^2 + k^2}$, but the number of eigenstates corresponding to a given eigenvalue is twice as big in the absence of the BC.[7] Of course these considerations are elementary, but they are crucial when trying to calculate (or just to give sense to) the Casimir energy density and force. More to this, just in the same way as the traces of the two matrices $M_1 = \text{diag}(\alpha, \beta)$ and $M_2 = \text{diag}(\alpha, \alpha, \beta, \beta)$ are not equal in spite of having "the

---

[7]To understand this point even better, consider the fact that further, by imposing Cauchy BC: $\phi(0) = 0, \phi'(0) = 0$, the eigenvalues still remain the *same*, but for any k the family of eigenfunctions shrinks to just the trivial one: $\phi_k(x) = 0, \forall k$ (the Cauchy problem is an initial value problem, which completely determines the solution).

same spectrum α, β," in the problem under discussion the traces of the Hamiltonian with and without the Dirichlet BC imposed yield different results, both of them divergent, namely

$$\operatorname{tr} H = 2 \operatorname{tr} H^{BC} = 2 \int_0^\infty dk \sqrt{m^2 + k^2}. \tag{7.49}$$

By using the zeta function, we define

$$\zeta^{BC}(s) := \int_0^\infty d\kappa \left(\nu^2 + \kappa^2\right)^{-s}, \quad \nu := \frac{m}{\mu}, \tag{7.50}$$

with μ a regularization parameter with dimensions of mass.[8] We get

$$\zeta^{BC}(s) = \frac{\sqrt{\pi}\,\Gamma(s - 1/2)}{2\Gamma(s)} \left(\nu^2\right)^{1/2-s}, \tag{7.51}$$

and consequently,

$$\operatorname{tr} H^{BC} = \frac{1}{2}\zeta_{BC}(s = -1/2)$$

$$= \frac{m^2}{4\sqrt{\pi}} \left[\frac{1}{s + 1/2} + 1 - \gamma - \log\frac{m^2}{\mu^2} - \Psi(-1/2) + \mathcal{O}(s + 1/2) \right]\Bigg|_{s=-1/2}. \tag{7.52}$$

As is obvious, this divergence is *not cured* when taking the difference of the two traces in order to obtain the Casimir energy:

$$E_C/\mu = E_0^{BC}/\mu - E_0/\mu = -E_0^{BC}/\mu = \frac{\Gamma(-1)m^2}{8\mu^2}. \tag{7.53}$$

We just hit the pole of the zeta function, in this case. How may this infinity be interpreted? It clearly originates from the fact that imposing the BC has drastically reduced to one-half the family of eigenfunctions corresponding to any of the eigenvalues which constitute the spectrum of the operator. And we can also advance that, since this dramatic reduction of the family of eigenfunctions takes place precisely at the point where the BC is imposed, the physical divergence (infinite energy) will originate right there.

   While the analysis above cannot be taken as a substitute for the actual description of Jaffe et al. [215–218]—where the BC is explicitly enforced through the introduction of an auxiliary, localized field, which probes what happens at the boundary in a much more precise way—it certainly shows that *pure mathematical considerations,*

---

[8]Always necessary in zeta regularization, since the complex powers of the spectrum of a (pseudo-)differential operator can only be defined, physically, if the operator is rendered dimensionless, what is done by introducing this parameter. That is also an important issue (sometimes overlooked).

which include the use of analytic continuation by means of the zeta function, are not blind to the infinities of the physical model and do not necessarily produce misleading results, when the mathematics are used properly. It is remarkable to realize how close the mathematical description of the appearance of an infinite contribution is to the one provided by the physical model in [215–218].

7.3.2 Case of Two-Point Dirichlet Boundary Conditions

A similar analysis can be done for the case of a two-point Dirichlet BC: $\phi(a) = 0$, $\phi(-a) = 0$. Now the eigenvalues k are quantized, as $k = \dot{\pi}/(2a)$, so that:

$$\omega_\ell = \sqrt{m^2 + \frac{\ell^2 \pi^2}{4a^2}}, \quad \ell = 0, 1, 2, \ldots. \tag{7.54}$$

The family of eigenfunctions corresponding to a given eigenvalue, ω_ℓ, is of continuous dimension 1, exactly as in the former case of a one-point Dirichlet BC, namely, $\phi_\ell(x) = b \sin(\frac{\ell\pi}{2a}(x-a))$, where b is an arbitrary, real parameter.[9] To repeat, the act of imposing Dirichlet BC on two points has the effect of discretizing the spectrum but there is no further shrinking in the number of eigenfunctions corresponding to a given (discrete) eigenvalue.

The calculation of the Casimir energy, by means of the zeta function, proceeds in this case as follows [10, 14, 204–212, 222–226]. It may be interesting to recall that the zeta-'measure' of the continuum equals twice the zeta-'measure' of the discrete. In fact, just consider the following regularizations: $\sum_{n=1}^\infty \mu = \mu \sum_{n=1}^\infty n^{-s}|_{s=0} = \mu \zeta_R(0) = -\frac{\mu}{2}$, and $\int_\mu^\infty dk = \int_0^\infty dk \, (k+\mu)^{-s}|_{s=0} = \frac{\mu^{1-s}}{s-1}|_{s=0} = -\mu$, which prove the statement.

The trace of the Hamiltonian corresponding to the quantum system with the BC imposed, in the massive case, is obtained by means of the zeta function

$$\zeta^{BC}(s) := \sum_{\ell=1}^\infty \left(\frac{m^2}{\mu^2} + \frac{\pi^2 \ell^2}{4\mu^2 a^2} \right)^{-s}$$

$$= \left(\frac{\mu}{m} \right)^{2s} \left[-\frac{1}{2} + \frac{\Gamma(s-1/2)}{\Gamma(s)} \frac{am}{\sqrt{\pi}} \right.$$

$$\left. + \frac{2\pi^s}{\Gamma(s)} \left(\frac{2am}{\pi} \right)^{1/2+s} \sum_{n=1}^\infty n^{s-1/2} K_{s-1/2}(4anm) \right], \tag{7.55}$$

---

[9]The contribution of the zero-mode ($\ell = 0$) is controverted, but we are not going to discuss this issue here (see e.g. [227] and references therein).

being K_ν a modified Bessel function of the third kind (or MacDonald's function). Thus, for the zero point energy of the system with two-point Dirichlet BC, we get

$$\operatorname{tr} H^{BC}/\mu = \frac{1}{2}\zeta_{BC}(s=-1/2) = -\frac{\Gamma(-1)m^2}{8\mu^2} - \frac{m}{2\pi\mu}\sum_{n=1}^{\infty}\frac{1}{n}K_1(2\pi nm/\mu), \quad (7.56)$$

where μ is, in this case, $\mu := \pi/(2a)$ (a fixes the mass scale in a natural way here). As in the previous example, one finally obtains an infinite value for the Casimir energy, namely

$$E_C/\mu = E_0^{BC}/\mu - E_0/\mu = \frac{\Gamma(-1)m^2}{8\mu^2} - \frac{m}{2\pi\mu}\sum_{n=1}^{\infty}\frac{1}{n}K_1(2\pi nm/\mu). \quad (7.57)$$

It is, therefore, not true that regularization methods using analytical continuation (in particular, the zeta approach) are unable to see the infinite energy that is generated on the boundary-conditions surface [213–218] (see (7.66) later). The reason is again, as in the previous example, that imposing a two-point Dirichlet BC amounts to halving the family of eigenfunctions which correspond to any given eigenvalue (all are discrete, in the present case, but this makes no difference). In physical terms, that means having to apply an infinite amount of energy on the BC sites, in order to enforce the BC. Analogously, from the mathematical viewpoint, halving the family of eigenfunctions results in the appearance of an infinite contribution, under the form of a pole of the zeta function.

The reason why these infinities (the one here and that in the previous section) do not usually show up in the literature on the Casimir effect is probably because textbooks on the subject focus towards the calculation of the Casimir *force*, which is obtained by taking minus the derivative of the energy with respect to the plate (or point) separation (here w.r.t. $2a$). Since the infinite terms do not depend on a, they *do not* contribute to the force (see also [215–218]). However, some misleading statements may have appeared in some classical references, stemming from the lack of recognition of the catastrophical implications of the act of halving the number of eigenfunctions, when imposing the BC.

7.3.3 How to Deal with the Infinities?

The infinite contributions have here shown up at the regularization level, but according to [215–218] they can persist even after renormalizing in a proper way. The important question is: are these infinities *physical*? Will they be *observed* as a manifestation of a very large energy pressure when approaching the BC surface in a lab experiment? Such questions will be only answered experimentally, and up to now there is no trace of them. If, on the contrary, as those large pressures fail to become manifest this is an indication of the need for an additional regularization prescription. In principle, this seems to be forbidden by standard renormalization theory, since the procedure has been already carried out to the very end (see [215–218]).

There are circumstances—both in physics and in mathematics—where other regularization methods have been employed with good success. In particular, Hadamard regularization in higher-post-Newtonian general relativity [228], and in recent variants of axiomatic and constructive QFT [229–234]. Among mathematicians, Hadamard regularization is nowadays a rather standard technique to deal with singular differential and integral equations with BCs, both analytically and numerically (for a sample of references see [235–240]). Indeed, Hadamard regularization is a well-established procedure in order to give sense to infinite integrals. It is not to be found in the classical books on infinite calculus by Hardy or Knopp; it was L. Schwartz [241] who popularized it, rescuing Hadamard's original papers. Nowadays, Hadamard convergence is one of the cornerstones in the rigorous formulation of QFT through micro-localization, which on its turn is considered by specialists to be the most important step towards the understanding of linear PDEs since the invention of distributions [242] (for a beautiful treatment of Hadamard's regularization see [243]).

Let us briefly recall this formulation. Consider a function, $g(x)$, expandable as

$$g(x) = \sum_{j=1}^{k} \frac{a_j}{(x-a)^{\lambda_j}} + h(x), \tag{7.58}$$

with λ_j complex in general and $h(x)$ a regular function. Then, it is immediate that $\int_{a+\epsilon}^{b} dx\, g(x) = P(1/\epsilon) + H(\epsilon)$, being P a polynomial and $H(0)$ finite. If the $\lambda_j \notin \mathbb{N}$, then one defines the Hadamard regularized integral as

$$\fint_{a}^{b} dx\, g(x) := \int_{a}^{b} h(x)\, dx - \sum_{j=1}^{k} \frac{a_j}{\lambda_j - 1}(b-a)^{1-\lambda_j}. \tag{7.59}$$

Alternatively, one may define, for $\alpha \notin \mathbb{N}$, $p < \alpha < p+1$, and $f^{(p+1)} \in C_{[-1,1]}$, $K^\alpha f := \frac{1}{\Gamma(-\alpha)} \fint_{-1}^{1} dt\, \frac{f(t)}{(1-t)^{\alpha+1}}$, to obtain, after some steps,

$$K^\alpha f = \sum_{j=0}^{p} \frac{f^{(j)}(-1)}{\Gamma(j+1-\alpha)2^{\alpha-j}} + \frac{1}{\Gamma(p+1-\alpha)}$$

$$- \int_{-1}^{1} (1-t)^{p-\alpha} f^{(p+1)}(t), \tag{7.60}$$

where the last integral is at worst improper (Cauchy's principal part). If $\lambda_1 = 1$, the result is $a_1 \ln(b-a)$, instead. If $\lambda_1 = p \in \mathbb{N}$, calling $H_p(f; x) := \fint_{-1}^{1} dt\, \frac{f(t)}{(t-x)^{p+1}}$, $|x| < 1$, we get

$$H_p(f; x) = \int_{-1}^{1} \left[f(t) - \sum_{j=0}^{p} \frac{f^{(j)}(x)}{j!} (t-x)^j \right] \frac{dt}{(t-x)^{p+1}} + \frac{f^{(j)}(x)}{j!}$$

$$= \int_{-1}^{1} \frac{dt}{(t-x)^{p+1-j}}, \tag{7.61}$$

where the first term is regular and the second one can be easily reduced to

$$\frac{1}{(p-j)!} \frac{d^{p-j}}{dx^{p-j}} - \int_{-1}^{1} \frac{dt}{t-x}, \tag{7.62}$$

being the last integral, as before, a Cauchy PP.

An alternative form of Hadamard's regularization, which is more fashionable for physical applications (as is apparent from the expression itself) is the following [228]. For the case of two singularities, at $\vec{x}_1$, $\vec{x}_2$, after excising from space two little balls around them, $\mathbb{R}^3 \backslash (B_{r_1}(\vec{x}_1) \cup B_{r_2}(\vec{x}_2))$, with $B_{r_1}(\vec{x}_1) \cap B_{r_2}(\vec{x}_2) = \emptyset$, one defines the regularized integral as being the finite part of the limit

$$\fint d^3x \, F(\vec{x}) := \mathrm{FP}_{\alpha,\beta \to 0} \int d^3x \left(\frac{r_1}{s_1}\right)^\alpha \left(\frac{r_2}{s_2}\right)^\beta F(\vec{x}), \tag{7.63}$$

where s_1 and s_2 are two (dimensionfull) regularization parameters [228]. This is the version that will be employed in what follows.

7.3.4 Hadamard Regularization of the Casimir Effect

We now use Hadamard's regularization as an additional tool in order to make sense of the infinite expressions encountered in the boundary value problems considered before. As it turns out from a detailed analysis of the results in [215–218], the basic integrals which produce infinities, in the one-dimensional and two-dimensional cases there considered, are the following.

In one dimension, with Dirichlet BC imposed at one ($x = 0$) and two ($x = \pm a$) points, respectively, by means of a delta-background of strength λ (see [215–218]), one encounters the two divergent integrals:

$$E_1(\lambda, m) = \frac{1}{2\pi} \int_m^\infty \frac{dt}{\sqrt{t^2 - m^2}} \left[t \log\left(1 + \frac{\lambda}{2t}\right) - \frac{\lambda}{2} \right], \tag{7.64}$$

$$E_2(a, \lambda, m) = \frac{1}{2\pi} \int_m^\infty \frac{dt}{\sqrt{t^2 - m^2}} \left\{ t \log\left[1 + \frac{\lambda}{t} + \frac{\lambda^2}{4t^2}(1 - e^{-4at}) \right] - \lambda \right\}. \tag{7.65}$$

Using Hadamard's regularization, as described before, we obtain for the first one, (7.64),

$$E_1(m) = \frac{\lambda}{4\pi} \left(1 - \ln \frac{\lambda}{m}\right) \Big|_{\lambda \to \infty} + \fint, \tag{7.66}$$

where the first term is the singular part when the limit $\lambda \to \infty$ is taken, and the second—which is Hadamard's finite part—yields in this case

$$\fint = -\frac{m}{4}. \tag{7.67}$$

Such result is coinciding with the classical one (0, for $m = 0$). Note in particular, that the further $\ln m$ divergence as $m \to \infty$ is hidden in the λ-divergent part, and that behavior does explain why the classical results which are obtained using hard Dirichlet BC—what corresponds as we just prove here to the Hadamard's regularized part—cannot see it.

In the case of a two-point boundary at $x = \pm a$ (separation $2a$), (7.65), we get a similar equation (7.66) but now the regularized integral is as follows. For the massless case, we obtain

$$\fint = -\frac{\pi}{48a}, \tag{7.68}$$

which is the regularized result to be found in the classical books. In the massive case, $m \neq 0$, after some additional work the following fast convergent series turns up [cf. (7.57)]

$$\fint = -\frac{m}{2\pi} \sum_{k=1}^{\infty} \frac{1}{k} K_1(4akm). \tag{7.69}$$

Thus (7.66) yields strictly the same result (7.57) that was already obtained by imposing the Dirichlet BC *ab initio*. What has now been *gained* is a more clear identification of the singular part, in terms of the strength of the delta potential at the boundary. This will be the general conclusion, common to all the other cases here considered.

Correspondingly, for the Casimir force we obtain the finite values[10]

$$F_2(a) = -\frac{\pi}{96a^2}, \tag{7.70}$$

in the massless case, and in the massive one

$$F_2(a, m) = -\frac{m^2}{\pi} \sum_{k=1}^{\infty} \left[K_0(4akm) + \frac{1}{4akm} K_1(4akm) \right]. \tag{7.71}$$

Those expressions coincide with the ones derived in the above mentioned textbooks on the Casimir effect, and reproduced before by using the zeta function method (just take minus the derivative of (7.66) w.r.t. $2a$).

The two-dimensional case turns out to be more singular [215–218]—in part just for dimensional reasons—and requires additional wishful thinking in order to deal

---

[10]Note that the force $F(a)$ is here given by minus the derivative of the total energy $E(a)$ w.r.t. $2a$, since this is the distance between the two Dirichlet points (*not* a).

with the circular delta function sitting on the circumference where the Dirichlet BC are imposed. Here one encounters the basic singular integral, for the term contributing to the second Born approximation (we use the same notation as in [215–218]),

$$\tilde{\sigma}(p) = \int_0^\infty dr\, r J_0(pr)\sigma(r), \quad \sigma(r) = b\lambda \exp\left[-\frac{(r-a)^2}{2\omega^2}\right], \quad (7.72)$$

with J_0 a Bessel function of the first kind, and $\int_0^\infty dr\,\sigma(r) = \lambda$, $\sigma(r) \xrightarrow{\omega\to0} \lambda\delta(r-a)$. Hadamard's regularization yields now (the τ's replacing the σ's in the regularized version)

$$\tau(r, p) = c\lambda(rp + 1)^{-\omega/2} \exp\left[-\frac{(r-a)^2}{2\omega^2}\right] \xrightarrow{\omega\to0} \lambda\delta(r-a), \quad (7.73)$$

with p a (dimensionfull) regularization parameter, being the constant c given by $c^{-1} = \int_0^\infty dr\, r^{-\omega} \exp[-\frac{(r-a)^2}{2\omega^2}]$, which exists and is perfectly finite; in particular, $c^{-1}(\omega = .1, a = 1) = .25$. Then,

$$\tilde{\tau}(p) = 2\pi \int_0^\infty dr\, r J_0(pr)\tau(r, p) = 2\pi\lambda a(ap + 1)^{-\omega/2} J_0(ap) \quad (7.74)$$

It turns out that, for the Casimir energy, we get in this case (notation as in [215–218])

$$E_{\lambda^2}^{(2)}[\tau] = \frac{\lambda^2 a^2}{8} \int_0^\infty dp\, (ap + 1)^{-\omega} J_0(ap)^2 \arctan(p/2m)|_{\omega\to0}$$

$$= \frac{\lambda^2 a^2}{8}\left\{\frac{1}{2\omega} + \frac{\gamma + 3\ln 2}{2a} + 4m\left[\gamma - \frac{2}{\sqrt{\pi}}[1 - \ln(am)]h(4a^2m^2)\right]\right\}, \quad (7.75)$$

where $h(z) := {}_2F_3((1/2, 1/2); (1, 1, 3/2); z)$ and γ is the Euler–Mascheroni constant; in particular, for instance $h(1) = 1.186711$, what is quite a nice value. Recall also that ω is the *width* of the Gaussian δ, which is the very *physical* parameter considered in [213]. When this width tends to zero an infinite energy appears (the width controls the formation of the pole). The rest of the result is the Hadamard regularization of the integral, e.g.[11]

$$\fint_0^\infty dp\, J_0(ap)^2 \arctan(p/2m). \quad (7.76)$$

Again, the finite part reverts to the results obtained in the literature with Dirichlet BC *ab initio*.

---

[11]It should be pointed out that the computational program Mathematica [244] directly assigns the Hadamard regularized value to particular cases of integrals of this kind; but it does so without any hint on what is going on. This has often confused more one user, who fails to understand how it comes that an infinite integral gets a finite value out of nothing.

To summarize, we have here proven—in some particular but rather non-trivial and representative examples—that the finite results derived through the use of Hadamard's regularization exactly coincide with the values obtained using the more classical, less full-fledged methods to be found in the literature on the Casimir effect. Moreover, Hadamard's prescription is able to separate and identify the singularities as physically meaningful cut-offs. Although the validity of this additional regularization is at present questionable, the fact that it bridges the two approaches is already remarkable, maybe again a manifestation of the unreasonable effectiveness of mathematics.

Chapter 8
Applications to Gravity, Strings and p-Branes

In this chapter we are going to present two more examples of applications of the method of zeta function regularization in two different, although related, contexts. One of them is the spontaneous compactification—in the case of a $\mathbb{R}^1 \times \mathbb{S}^1$ background—that occurs in two-dimensional quantum gravity. This theory is considered to be an adequate toy model for a more fundamental formulation of quantum gravity. As an interesting physical result, with the help of the method of zeta-function regularization advocated here we will see [245] that such compactification is stable, in contradistinction to what happens in multidimensional quantum gravity on a $\mathbb{R}^d \times \mathbb{S}^1$ background (with $d > 2$), which is known to be one-loop unstable.

In Sect. 8.2 we obtain the effective potential for strings and p-branes, in general, and specifically study the stability of the rigid membrane [246]. A careful analysis of the ζ-functions relevant for the calculation of the effective potential for fixed-end and toroidal rigid p-branes at one-loop order and in the $1/d$ approximation—which are of inhomogeneous Epstein type—is performed. Asymptotic formulas which give accurate results (allowing only for exponentially decreasing errors of order $\mathcal{O}(10^{-3})$) are presented for the general case of p-branes, which carry all the dependencies on the basic parameters of the theory explicitly, and the behavior of the effective potential—specified to the membrane case ($p = 2$)—will be investigated. Finally, the extrema of this effective potential will be obtained.

8.1 Application to Spontaneous Compactification in Two-Dimensional Quantum Gravity

Let us consider induced two-dimensional gravity, with the action

$$S = \int d^2x \sqrt{g} \left(R \frac{1}{\Delta} R + \Lambda \right), \tag{8.1}$$

on the background $\mathbb{R}^1 \times \mathbb{S}^1$. On such a background—which is not the solution of the classical equations of motion—the convenient effective action is always gauge

E. Elizalde, *Ten Physical Applications of Spectral Zeta Functions*,
Lecture Notes in Physics 855,
DOI 10.1007/978-3-642-29405-1_8, © Springer-Verlag Berlin Heidelberg 2012

dependent. However, the S-matrix (the effective action *on shell*, i.e., at the stationary points) is independent of the choice of gauge condition. Actually, working in the loop expansion, one is led to an explicit gauge dependence even on shell (perturbatively). This is why it is preferable to work with the gauge-independent effective action.

Using the standard background field method,

$$g_{\mu\nu} \longrightarrow g_{\mu\nu} + h_{\mu\nu}, \qquad (8.2)$$

where $g_{\mu\nu}$ is the metric of flat space $\mathbb{R}^1 \times \mathbb{S}^1$ and $h_{\mu\nu}$ is the quantum gravitational field, choosing the gauge fixing action as

$$S_{GF} = \frac{1}{\alpha} \int d^2x \sqrt{g} \left(\nabla_\mu h^\mu_\rho - \beta \nabla_\rho h \right)^2, \qquad (8.3)$$

where α and β are the gauge parameters and $h = h^\mu_\mu$, and defining the configuration-space metric in accordance with Vilkoviski [247, 248]

$$\gamma_{ij} \equiv \gamma_{g_{\mu\alpha}g_{\nu\beta}} = \frac{1}{2}\sqrt{g}\left(g^{\mu\alpha}g^{\nu\beta} + g^{\mu\beta}g^{\nu\alpha} - ag^{\mu\nu}g^{\alpha\beta} \right), \qquad (8.4)$$

where a is a constant parameter, after some work, one obtains the following result for the one-loop effective action [245]

$$\Gamma^{(1)} = 2\pi R S\Lambda + \frac{1}{2}\left[\mathrm{tr}\ln\left(\Delta + \frac{\Lambda}{4(2-a)} \right) - 2\,\mathrm{tr}\ln\Delta \right]. \qquad (8.5)$$

Here $2\pi R$ is the length of the compactified dimension while $S = \int dx$ is the 'volume' of the space $\mathbb{R}^1$. As we see, the dependence on the gauge parameters α and β has disappeared. However, an explicit dependence on the parameter a remains.

The trace calculations involved in expression (8.5) for the one-loop effective action are not easy to perform. The usual non-trivial commutations of series have to be carried out. Using the formulas of Chap. 4 (specified to $\mathbb{R}^1 \times \mathbb{S}^1$), we obtain

$$\zeta_{-\Delta+m^2}\left(\frac{s}{2}\right) = -S \int_0^\infty \frac{dk}{\pi} \sum_{n=-\infty}^{+\infty} \left[k^2 + \left(\frac{2\pi n}{\beta}\right) + m^2 \right]^{-s/2}$$

$$= -\frac{S}{\sqrt{\pi}} m^{1-s} \left\{ \frac{-\Gamma(\frac{s-1}{2})}{2\Gamma(\frac{s}{2})} + \frac{\beta m}{2\sqrt{\pi}} \frac{1}{s-2} + \frac{(\frac{\beta m}{2})^{(s-1)/2}}{\Gamma(\frac{s}{2})} \sum_{k=0}^{\infty} \frac{(16\pi)^{-k}}{k!} \right.$$

$$\left. \cdot \left(\frac{2\pi}{\beta m}\right)^k \prod_{j=1}^{k} [(s-2)^2 - (2j-1)^2] \sum_{n=1}^{\infty} n^{\frac{s-3}{2}-k} e^{-\beta mn} \right\}, \qquad (8.6)$$

wherefrom we get

$$V = \frac{\Gamma^{(1)}}{S} = 2\pi R\Lambda + \frac{R\Lambda}{32(2-a)}\left[1 - \ln\left(\frac{\Lambda}{4(2-a)}\right) \right] - \frac{1}{8}\sqrt{\frac{\Lambda}{2-a}} + \frac{1}{24R}$$

$$-\frac{1}{4\pi\sqrt{2R}}\left(\frac{\Lambda}{2-a}\right)^{1/4}\sum_{k=0}^{\infty}\frac{(16\pi)^{-k}}{k!}\left(\frac{R}{2}\sqrt{\frac{\Lambda}{2-a}}\right)^{-k}$$

$$\cdot\prod_{j=1}^{k}[4-(2j-1)^2]\sum_{n=1}^{\infty}n^{-(k+3/2)}\exp\left(-\pi R\sqrt{\frac{\Lambda}{2-a}}n\right). \qquad (8.7)$$

This expression can be now simplified in terms of the basic variables of the problem:

$$x\equiv\frac{\Lambda}{4(2-a)}, \qquad y\equiv R\sqrt{x}=\frac{R}{2}\sqrt{\frac{\Lambda}{2-a}}. \qquad (8.8)$$

Then, the effective action becomes

$$V=\sqrt{x}\left[8\pi(2-a)y+\frac{y}{8}(1-\ln x)-\frac{1}{4}+\frac{1}{24y}-F(y)\right], \qquad (8.9)$$

$F(y)$ being given by

$$F(y)=\frac{1}{4\pi}\sum_{k=0}^{\infty}\frac{(16\pi)^{-k}}{k!}y^{-(k+1/2)}\prod_{j=1}^{k}[4-(2j-1)^2]\sum_{n=1}^{\infty}n^{-(k+3/2)}e^{-2\pi ny}.$$

$$(8.10)$$

It is now clear that all the dependence of the action on R, Λ and a comes through the specific combination given by the variable y, but for a global factor, $\sqrt{x}$, and for the first term, which is just linear in a.

To proceed with the compactification program, one imposes (as is done in multi-dimensional gravity)

$$\begin{cases} V(R,\Lambda,a)=0, \\ \dfrac{\partial V(R,\Lambda,a)}{\partial R}=0. \end{cases} \qquad (8.11)$$

The explicit a dependence can be eliminated, and one gets

$$\sqrt{x}\left[F(y)-yF'(y)-\frac{1}{12y}+\frac{1}{4}\right]=0. \qquad (8.12)$$

This transcendent equation involves an asymptotic series, and must be solved approximately. By fortune, the decreasing exponentials come to rescue and, after an explicit calculation one obtains the (expected) result:

$$y_1=0.33. \qquad (8.13)$$

This is the non-trivial stationary point of the effective action. The trivial one is reached for

$$x_0=0. \qquad (8.14)$$

As for the second derivative,

$$\frac{\partial^2 V}{\partial y^2} = \sqrt{x}\left[\frac{1}{12y^3} - F''(y)\right], \tag{8.15}$$

where the explicit a-dependence has disappeared. Hence, this derivative has a definite sign (independent of a) at the stationary point

$$\left.\frac{\partial^2 V}{\partial y^2}\right|_{y=y_1} \simeq 2 > 0. \tag{8.16}$$

The point is a minimum, obtained for the following combination of parameters

$$\frac{\Lambda \mathbb{R}^2}{2-a} \simeq \left(\frac{2}{3}\right)^2. \tag{8.17}$$

We see that the compactification is stable, in contradistinction to what happens in multidimensional quantum gravity on a $\mathbb{R}^d \times \mathbb{S}^1$ background, for $d > 2$, which is known to be one-loop unstable.

8.2 Application to the Study of the Stability of the Rigid Membrane

The theory of the rigid string is interesting because of the number of its applications to quantum chromodynamics [249–256] and to statistical physics. Using the same idea, it is not difficult to construct the action for the rigid membrane (and p-brane in general). During the last few years, there has been some activity in the study of quantum extended objects [257–259]. However, already the semiclassical quantization of such a nonlinear system as a membrane is a difficult task [260–262]. Nevertheless, some interesting issues, like the study of the Casimir energy, the large-d approximation and the tachyon problem can be addressed already at the semiclassical level [263–266].

In the present section, we will study the Casimir energy and the static potential for the rigid p-brane (at the classical level, this theory has been considered in Ref. [267]), specifying afterwards our results to the membrane case, $p = 2$. We shall start from the following action, which is multiplicatively renormalized only in the string case ($p = 1$),

$$S = \int d^{p+1}\xi \sqrt{g}\left(k + \frac{1}{2\rho^2}[\Delta(g)X^i]^2\right), \tag{8.18}$$

where $g_{\alpha\beta} = \partial_\alpha X^i \partial_\beta X^i$, $\alpha = 0, 1, \ldots, p$, $i = 1, 2, \ldots, d$, $\Delta(g) = g^{-1/2}\partial_\alpha g^{1/2} \times g^{\alpha\beta}\partial_\beta$, the constant k is the analog of the usual string tension, and $1/\rho^2$ is the coupling constant corresponding to the rigid term.

Let us note that the p-brane is an interacting system, without a free part in the action. Hence, one must start from some classical solution for the ground state, and then study the quantum fluctuations on such background. In this framework we can understand how the tachyon appears (if that is the case), if the background is stable, and also address some other issues. Owing to the fact that string theory can be obtained as some compactification of the membrane [268], we can also expect to find in this way some new features of string physics.

In Sect. 8.2.1 we calculate the potential corresponding to two cases: fixed-end and periodic boundary conditions. In Sect. 8.2.2 we obtain the static potential, that is, the effective potential in the limit of large spacetime dimensionality. Owing to the difficulty of the exact expressions, a saddle point analysis is carried out in Sect. 8.2.3. In Sect. 8.2.4 we apply the useful mathematical results on the inhomogeneous Epstein-type zeta functions obtained in Chap. 3 to the expressions that appear in the process of regularization. Finally, in Sect. 8.2.5 we present a short discussion of the general case.

8.2.1 Calculation of the Potential

We consider as the background the classical solutions of the field equations [263–266] (which are the same for the rigid as for the usual p-brane)

$$X_{cl}^0 = \xi_0, \qquad X_{cl}^\perp = 0, \qquad X_{cl}^{d-1} = \xi_1, \quad \ldots, \quad X_{cl}^{d-p} = \xi_p, \qquad (8.19)$$

with $X_{cl}^\perp = (X^1, \ldots, X^{d-p-1})$ and $(\xi_1, \ldots, \xi_p) \in \mathcal{R} \equiv [0, a_1] \times \cdots \times [0, a_p]$. We also use the axial gauge

$$X^0 = X_{cl}^0, \qquad X^{d-1} = X_{cl}^{d-1}, \quad \ldots, \quad X^{d-p} = X_{cl}^{d-p}, \qquad (8.20)$$

where the Faddeev–Popov ghosts are absent.

We shall consider the toroidal rigid p-brane which has the boundary conditions

$$X^\perp(0, \xi_1, \ldots, \xi_p) = X^\perp(T, \xi_1, \ldots, \xi_p) = 0 \qquad (8.21)$$

and

$$X^\perp(\xi_0, 0, \xi_2, \ldots, \xi_p) = X^\perp(\xi_0, a_1, \xi_2, \ldots, \xi_p),$$

$$\vdots \qquad\qquad (8.22)$$

$$X^\perp(\xi_0, \xi_1, \ldots, \xi_{p-1}, 0) = X^\perp(\xi_0, \xi_1, \ldots, \xi_{p-1}, a_p).$$

For the fixed-end boundary conditions, (8.21) is exactly the same, while (8.22) are replaced by the following (of Dirichlet type)

$$X^\perp(\xi_0, 0, \xi_2, \ldots, \xi_p) = \cdots = X^\perp(\xi_0, \xi_1, \ldots, \xi_{p-1}, 0),$$

$$X^\perp(\xi_0, a_1, \xi_2, \ldots, \xi_p) = \cdots = X^\perp(\xi_0, \xi_1, \ldots, \xi_{p-1}, a_p). \qquad (8.23)$$

The effective potential is given by

$$V = - \lim_{T \to \infty} \frac{1}{T} \ln \int \mathcal{D}X^\perp \exp(-S).$$
(8.24)

Restricting ourselves to the one-loop approximation, we need only consider the terms which are quadratic in the quantum fields (this applies to the usual membrane and p-brane, see [263–266]).

Integrating out $X^\perp$ and using boundary conditions to read off the resulting $\mathrm{Tr} \ln \Delta$ (see [263–266, 269, 270]), we get

$$V_{fixed\ end} = k \prod_{i=1}^{p} a_i + \frac{d-p-1}{2} \left[\sum_{n_1,\ldots,n_p=1}^{\infty} \left(\frac{\pi^2 n_1^2}{a_1^2} + \cdots + \frac{\pi^2 n_p^2}{a_p^2} \right)^{1/2} \right.$$
$$\left. + \sum_{n_1,\ldots,n_p=1}^{\infty} \left(\frac{\pi^2 n_1^2}{a_1^2} + \cdots + \frac{\pi^2 n_p^2}{a_p^2} + k\rho^2 \right)^{1/2} \right],$$
(8.25)

and

$$V_{toroidal} = k \prod_{i=1}^{p} a_i + \frac{d-p-1}{2} \left[\sum_{n_1,\ldots,n_p=-\infty}^{\infty} \left(\frac{4\pi^2 n_1^2}{a_1^2} + \cdots + \frac{4\pi^2 n_p^2}{a_p^2} \right)^{1/2} \right.$$
$$\left. + \sum_{n_1,\ldots,n_p=-\infty}^{\infty} \left(\frac{4\pi^2 n_1^2}{a_1^2} + \cdots + \frac{4\pi^2 n_p^2}{a_p^2} + k\rho^2 \right)^{1/2} \right].$$
(8.26)

Observe that the contribution from the higher-derivative mode appears in (8.25) and (8.26) with a positive sign, as it follows from the path integral.

8.2.2 The Limit of Large Spacetime Dimensionality

We calculate first the static potential—that is, the effective potential in the limit of large spacetime dimensionality. Such calculation for the usual Nambu–Goto or Eguchi string [271–273] has been carried out in Ref. [274] and for the rigid string in [252–255]. Let us introduce the composite fields $\sigma_{\alpha\beta}$ for $\partial_\alpha X^\perp \cdot \partial_\beta X^\perp$, and constrain $\sigma_{\alpha\beta} = \partial_\alpha X^\perp \cdot \partial_\beta X^\perp$ by introducing Lagrange multipliers $\lambda^{\alpha\beta}$:

$$Z = \int \mathcal{D}X^\perp \mathcal{D}\sigma \mathcal{D}\lambda \exp\left\{ -k \int d^{p+1}\xi \left[\det(\delta_{\alpha\beta} + \sigma_{\alpha\beta}) \right]^{1/2} \right.$$
$$\left. + \frac{1}{2\rho^2} \int d^{p+1}\xi \Delta_0 X^\perp \cdot \Delta_0 X^\perp - \frac{k}{2} \int d^{p+1}\xi \lambda^{\alpha\beta} \left(\partial_\alpha X^\perp \cdot \partial_\beta X^\perp - \sigma_{\alpha\beta} \right) \right\},$$
(8.27)

where $\Delta_0 = \eta^{\alpha\beta}\partial_\alpha\partial_\beta$. Integrating over $X^\perp$, we get

$$Z = \int \mathcal{D}\sigma \, \mathcal{D}\lambda \, \exp(-S_{\text{eff}}), \tag{8.28}$$

with

$$S_{\text{eff}} = \frac{1}{2}(d - p - 1)\operatorname{Tr}\ln\left(\frac{1}{\rho^2}\Delta_0^2 + k\partial_\alpha \lambda^{\alpha\beta}\partial_\beta\right)$$

$$+ kT\mathbb{R}^p\left[(1 + \sigma_0)^{1/2}(1 + \sigma_1)^{p/2} - \frac{1}{2}(\sigma_0\lambda_0 + p\sigma_1\lambda_1)\right], \tag{8.29}$$

where we have chosen $a_1 = \cdots = a_p = R$, and $\sigma_{\alpha\beta} = \operatorname{diag}(\sigma_0, \sigma_1, \ldots, \sigma_1)$, $\lambda_{\alpha\beta} = \operatorname{diag}(\lambda_0, \lambda_1, \ldots, \lambda_1)$ (compare with [263–266], where the case $1/\rho^2 = 0$ was considered).

In this case we obtain

$$\operatorname{Tr}\ln\left[\Delta_0^2 + k\rho^2\left(\lambda_0\partial_0^2 + \lambda_1\vec{\partial}_x^2\right)\right]$$

$$= \operatorname{Tr}\ln\left[\left(\partial_0^2 + \vec{\partial}_x^2\right)^2 + k\rho^2\left(\lambda_0\partial_0^2 + \lambda_1\vec{\partial}_x^2\right)\right]$$

$$\cdot \operatorname{Tr}\ln\left\{\left[\partial_0^2 + \left(\vec{\partial}_x^2 + \frac{k\rho^2\lambda_0}{2}\right) - \sqrt{k\rho^2(\lambda_0 - \lambda_1)\vec{\partial}_x^2 + \frac{k^2\rho^4\lambda_0^2}{4}}\right]\right.$$

$$\cdot\left.\left[\partial_0^2 + \left(\vec{\partial}_x^2 + \frac{k\rho^2\lambda_0}{2}\right) + \sqrt{k\rho^2(\lambda_0 - \lambda_1)\vec{\partial}_x^2 + \frac{k^2\rho^4\lambda_0^2}{4}}\right]\right\}. \tag{8.30}$$

The spectrum for the boundary conditions (8.19)–(8.23) is known. Using this spectrum and evaluating the Tr ln terms by means of analytic regularization for large T, we obtain (see [263–266] for details of this method)

$$S_{\text{eff}} = kT\mathbb{R}^p\left\{(1 + \sigma_0)^{1/2}(1 + \sigma_1)^{p/2} - \frac{1}{2}(\sigma_0\lambda_0 + p\sigma_1\lambda_1)\right.$$

$$+ \frac{d - p - 1}{2k\mathbb{R}^{p+1}}\left[\sum_{\vec{n}}\left(\pi^2\vec{n}^2 + \frac{k\rho^2\lambda_0\mathbb{R}^2}{2}\right.\right.$$

$$- \sqrt{k\rho^2(\lambda_0 - \lambda_1)\vec{n}^2\mathbb{R}^2 + \frac{k^2\rho^4\lambda_0^2\mathbb{R}^4}{4}}\,\right)^{1/2}$$

$$+ \sum_{\vec{n}}\left(\pi^2\vec{n}^2 + \frac{k\rho^2\lambda_0\mathbb{R}^2}{2} + \sqrt{k\rho^2(\lambda_0 - \lambda_1)\vec{n}^2\mathbb{R}^2 + \frac{k^2\rho^4\lambda_0^2\mathbb{R}^4}{4}}\,\right)^{1/2}\left.\left.\right]\right\}, \tag{8.31}$$

where for the fixed-end p-brane $\vec{n}^2 = n_1^2 + \cdots + n_p^2$ and $\sum_{\vec{n}}$ means $\sum_{n_1,\ldots,n_p=1}^{\infty}$, as in (8.25), while for the toroidal p-brane $\vec{n}^2 = 4(n_1^2 + \cdots + n_p^2)$ and $\sum_{\vec{n}}$ means $\sum_{n_1,\ldots,n_p=-\infty}^{\infty}$, as in (8.26).

8.2.3 A Saddle Point Analysis

The functions that appear on the right hand side of (8.31) are rather complicated to analyze. To our knowledge, they have never been considered in the literature and will be the object of a separate investigation. (Note that in the case of the usual Dirac p-brane [263–266] these functions are simply constants.) So we shall have here little to say about the corresponding effective potential, only, for example, that as $R \to \infty$, $V \sim V_{cl} = k\mathbb{R}^p$. Rewriting the expression for the static potential identically as

$$
V = k\mathbb{R}^p \left[(1+\sigma_0)^{1/2}(1+\sigma_1)^{p/2} - \frac{1}{2}(\sigma_0\lambda_0 + p\sigma_1\lambda_1) \right.
$$

$$
\left. + \frac{d-p-1}{2k\mathbb{R}^{p+1}} K\left(k\rho^2\mathbb{R}^2, \lambda_0, \lambda_1\right) \right],
\tag{8.32}
$$

we are led to the four saddle-point equations:

$$
\lambda_0 = (1+\sigma_1)^{p/2}(1+\sigma_0)^{-1/2},
$$

$$
\lambda_1 = (1+\sigma_1)^{p/2-1}(1+\sigma_0)^{1/2},
$$

$$
\sigma_0 = \frac{d-p-1}{k\mathbb{R}^{p+1}} \frac{\partial K(k\rho^2\mathbb{R}^2, \lambda_0, \lambda_1)}{\partial \lambda_0},
\tag{8.33}
$$

$$
\sigma_1 = \frac{d-p-1}{kp\mathbb{R}^{p+1}} \frac{\partial K(k\rho^2\mathbb{R}^2, \lambda_0, \lambda_1)}{\partial \lambda_1},
$$

By eliminating from these equations σ_0 and σ_1, we get

$$
(\lambda_1\lambda_0)^{p/(p-1)}\lambda_0^{-2} - 1 = \frac{d-p-1}{k\mathbb{R}^{p+1}} \frac{\partial K(k\rho^2\mathbb{R}^2, \lambda_0, \lambda_1)}{\partial \lambda_0},
$$

$$
(\lambda_1\lambda_0)^{1/(p-1)} - 1 = \frac{d-p-1}{kp\mathbb{R}^{p+1}} \frac{\partial K(k\rho^2\mathbb{R}^2, \lambda_0, \lambda_1)}{\partial \lambda_1}.
\tag{8.34}
$$

In principle, if not analytically, it is of course possible to eliminate (let us say) λ_0 from these two equations by means of a numerical calculation, and to rewrite V in terms of λ_1 only. After doing this, one can study V as a function of R and in terms of this parameter λ_1, in order to see if the tachyon is in the spectrum. We will not go through this here, since our purpose is just to show the possibility, in principle, of calculating the static potential for a rigid string by using the zeta-function method, in a quite simple way. However, we are still ready to look to some interesting limiting case of these equations. Let us assume that $\lambda_0 = \lambda_1 \equiv \lambda$ (such a choice has been taken for the rigid string in [252–255]). Then

$$
V = k\mathbb{R}^p \left\{ (1+\sigma_0)^{1/2}(1+\sigma_1)^{p/2} - \frac{1}{2}\lambda(\sigma_0 + p\sigma_1) \right.
$$
$$
\left. + \frac{(d-p-1)\pi}{2k\mathbb{R}^{p+1}} \left[\sum_{\vec{n}} \sqrt{\vec{n}^2} + \sum_{\vec{n}} (\vec{n}^2 + k(\rho/\pi)^2 \lambda \mathbb{R}^2)^{1/2} \right] \right\}. \quad (8.35)
$$

It follows from the saddle point equations that $\sigma_0 = \sigma_1$, $\lambda = (1+\sigma)^{(p-1)/2}$, and

$$
V = k\mathbb{R}^p \left[\lambda^{(p+1)/(p-1)} - \frac{p+1}{2}\lambda(\lambda^{2/(p-1)} - 1) + \frac{(d-p-1)\pi}{2k\mathbb{R}^{p+1}} K(k\rho^2\lambda\mathbb{R}^2) \right].
$$
$$
\quad (8.36)
$$

Here we have defined

$$
K(k\rho^2\lambda\mathbb{R}^2) \equiv \sum_{\vec{n}} \sqrt{\vec{n}^2} + \sum_{\vec{n}} (\vec{n}^2 + k(\rho/\pi)^2 \lambda\mathbb{R}^2)^{1/2}, \quad (8.37)
$$

and the last saddle point equation gives

$$
-\frac{p+1}{2}\lambda^{2/(p-1)} + \frac{(d-p-1)\pi}{2k\mathbb{R}^{p+1}} \frac{\partial K(k\rho^2\lambda\mathbb{R}^2)}{\partial\lambda} = 0. \quad (8.38)
$$

8.2.4 Explicit Expressions for the Zeta-Function Regularization of the Effective Potential

The expressions to be regularized are (in general) of the inhomogeneous Epstein form [51, 109, 275] (see Chaps. 3 and 4)

$$
E_p^c(s) \equiv \sum_{n_1,\dots,n_p=1}^{\infty} (n_1^2 + \cdots + n_p^2 + c^2)^{-s}, \quad (8.39)
$$

allowing for $c = 0$. These functions are not easy to deal with for $p > 1$ (let us repeat that the case $p = 1$ is the only one that has been investigated in the literature) and our results in Chaps. 2 and 4 are pioneering in this respect. There, general formulas have been derived (see (2.90)) for the functions

$$
M_N^c(s; a_1, \dots, a_N; \alpha_1, \dots, \alpha_N) \equiv \sum_{n_1,\dots,n_N=1}^{\infty} (a_1 n_1^{\alpha_1} + \cdots + a_N n_N^{\alpha_N} + c^2)^{-s}. \quad (8.40)
$$

In our case, $a_1 = \cdots = a_p = 1$ and $\alpha_1 = \cdots = \alpha_p = 2$, and (2.90) simplifies considerably (see (2.96) and (2.97)). Alternatively, useful recurrent formula for our case are (2.98) to (2.100) (see again Chap. 2)

In order to deal with the derivative of the function K above, one can follow two equivalent procedures: either do first the usual analytic continuation, and then take $s = -1/2$ and the derivative afterwards, or else take first the derivative of (8.39), perform the analytic continuation and put $s = +1/2$ at the end. The result is exactly the same. In either way, other non-trivial series commutations have to be performed (see Chap. 2, (2.100)).

As our final interest is numerical approximation (see above), we will not take into account exponentially small terms (let us point out that these expansions are asymptotic and quickly convergent, see Chap. 2). Notice, moreover, that for $c = 0$ there is no dependence on λ, so that the corresponding term does not contribute to (8.39).

From the mentioned expressions, for any value of p there is no difficulty in obtaining the value of λ which solves (8.39). In particular, for $p = 2$ we have:

$$\lambda \simeq \left(\frac{3-d}{12\pi}\right)^{2/3} k^{1/3}\rho^2. \tag{8.41}$$

This is an approximate root for $k \cdot \rho$ big (specifically, for $1/R \ll k^{2/3}\rho^2$) and $d \neq 3$. For $p = 3$ the result is:

$$\lambda \simeq \frac{1}{2^{10}}\left[\frac{3(4-d)}{\pi}\right]^2 \frac{k\rho^6}{\mathbb{R}^2}, \tag{8.42}$$

which is an approximate root for $k \cdot \rho$ big (specifically, for $k\rho^4 \gg 1$) and $d \neq 4$. Let us now substitute these values into the expression of V, and look for the derivative $\partial V/\partial R$. We get, for $p = 2$, an expression of the form

$$\frac{\partial V_2}{\partial R} = c_1 R + \frac{c_{-1}}{R} + \frac{c_{-2}}{\mathbb{R}^2}, \tag{8.43}$$

which has always one real root (at least), R_2. It corresponds to a minimum of V for $d > 3$ and reasonable values of the constants involved. Substituting back into $V(R)$ we see that the minimum is attained at

$$V_2(R_2) = kR_2^2\left[3(1+\sigma_0)^{1/2}(1+\sigma_1)\right.$$

$$\left. - \frac{3}{2}\left(\frac{3-d}{12\pi}\right)^{2/3}(\sigma_0 + 2\sigma_1)k^{1/3}\rho^2 + \frac{(d-3)^2}{48\pi^2}\rho^6\right], \tag{8.44}$$

i.e., for k big it is obtained for a negative value of V, while for ρ big it is reached for a positive value of V.

The case $p = 3$ is quite different. We get then

$$V_3(R) = \alpha_1\mathbb{R}^3 - \alpha_2 R + \frac{\beta}{R}, \tag{8.45}$$

so that its derivative has two real roots

$$R_\pm = \pm \left[\frac{\alpha_2 + (\alpha_2^2 + 12\alpha\beta)^{1/2}}{6\alpha_1} \right]^{1/2}, \tag{8.46}$$

one of which is seen to correspond to a maximum and the other to a minimum of V. Moreover, two additional, complex roots appear. The minimum for V is now attained at a negative value of V when either k or ρ are big enough and, conversely, at a positive value of the potential for k or ρ small. Note that in order to find the critical radius at which the potential becomes complex (so that the static approximation breaks down and tachyons appear) it is necessary to do the analysis with the general expression (8.32) directly. One can conjecture from our results here that the rigid membrane is tachyon-free (no critical radius exists), as it is also the case with rigid strings (see Refs. [253] and [256]). At least for the limiting situation discussed above, this is in fact the case.

8.2.5 Discussion of the General Case

As a summary, having done the calculation for this special case, corresponding to the limit of large spacetime dimensionality, and armed with the full equation (2.90), we can now be more ambitious and consider the one-loop effective potential, (8.25)-(8.26), without further limit or approximation. For the sake of conciseness, we shall restrict ourselves to $p = 2$—but it is obvious that we could consider as well any other value of p. We rely on equations (2.98), (2.99), and (2.101), (2.102), which we reproduce here because they are basic to our case:

$$\sum_{n_1,n_2=1}^{\infty} \sqrt{\left(\frac{n_1}{a_1}\right)^2 + \left(\frac{n_2}{a_2}\right)^2} = \frac{1}{24}\left(\frac{1}{a_1} + \frac{1}{a_2}\right) - \frac{\zeta(3)}{8\pi^2}\left(\frac{a_1}{a_2^2} + \frac{a_2}{a_1^2}\right)$$
$$- \frac{\pi^{3/2}}{2\sqrt{a_1 a_2}}\left[\exp\left(-2\pi\frac{a_1}{a_2}\right)\left(1 + \mathcal{O}(10^{-3})\right)\right], \tag{8.47}$$

and (this one after additional regularization, see above)

$$\sum_{n_1,n_2=1}^{\infty} \sqrt{\left(\frac{n_1}{a_1}\right)^2 + \left(\frac{n_2}{a_2}\right)^2 + c^2}$$
$$= \frac{c}{4} - \frac{\pi}{6}a_1 a_2 c^3 + \left(\frac{1}{4\pi}\sqrt{\frac{c}{a_2}} - \frac{ca_1}{4\pi a_2}\right)\left[\exp(-2\pi c a_2)\left(1 + \mathcal{O}(10^{-3})\right)\right]. \tag{8.48}$$

In both cases we have assumed (this is, of course, no restriction) that $a_2 \leq a_1$. These expressions are really valuable, the last term (already of exponential kind) being of order 10^{-3} with respect to the two first ones, and the not explicitly written contributions being of order 10^{-6} [246].

For fixed-end boundary conditions and not taking into account exponentially-small terms, we obtain

$$
V_{f.e.} \simeq ka_1 a_2 + \frac{(d-3)\pi}{24} \left[\frac{1}{2}\left(\frac{1}{a_1} + \frac{1}{a_2} \right) \right.
$$
$$
\left. - \frac{3\zeta(3)}{2\pi^2}\left(\frac{a_1}{a_2^2} + \frac{a_2}{a_1^2} \right) + \frac{3}{\pi}\sqrt{k}\rho - \frac{2}{\pi^2}k^{3/2}\rho^3 a_1 a_2 \right]. \quad (8.49)
$$

It is now straightforward to perform the analysis of extrema of V. For brevity, we shall only discuss here some particular cases. First, the one which is obtained from the two Lagrange equations for the extrema of V as a function of a_1 and a_2 only, for $a_1 = a_2 \equiv a$, which is reached for

$$
a = \left(\frac{\frac{3-d}{8}[3\zeta(3) - \pi^2]}{12\pi k + (3-d)k^{3/2}\rho^3} \right)^{1/3}. \quad (8.50)
$$

It can be seen that for $12\pi k + (3-d)k^{3/2}\rho^3 < 0$ this point is a minimum. On the contrary, it is a maximum for $12\pi k + (3-d)k^{3/2}\rho^3 > 0$. Consistency with the range of validity of the series expansion above is obtained for

$$
2\pi\sqrt{k}\rho > \frac{6}{\rho^2} \gg 1. \quad (8.51)
$$

typical values for which this is valid are: $\rho \simeq 2/3$, $k \simeq 4$, $2\pi c \simeq 8$.

Keeping now a_1 and a_2 fixed (but arbitrary), we see that (for $d > 3$) in terms of ρ, V is unbounded from below, being always negative for ρ big enough. Considering V as a function of k, the situation is similar. Finally, in a sense the analysis of Refs. [263, 264] is still valid here, when we fix the values of k, ρ and of the area $A = a_1 a_2$: the minima of the potential are obtained for elongated (stringy) membranes (a_1/a_2 small). Notice, however, that even for the particular case considered in [263, 264], our asymptotic expansion provides a more universal expression, because it is valid for any value of $a_2 \leq a_1$ (this is again not restrictive, in the end). It also goes without saying that, from our general formula (8.49) for the potential $V = V(a_1, a_2, k, \rho, d)$, one can perform a simultaneous analysis on *all* the different parameters at the same time—e.g. in order to look for local minima of the potential hypersurface—the explicit dependencies on k and ρ being also basic contributions of the present work.

In the case of toroidal boundary conditions, again neglecting exponentially-small contributions, we get (for a detailed discussion of the relations between the different boundary conditions, see Ref. [275])

$$
V_{tor} \simeq ka_1 a_2 + \frac{(d-3)\pi}{2} \left[-\frac{\zeta(3)}{\pi^2}\left(\frac{a_1}{a_2^2} + \frac{a_2}{a_1^2} \right)\frac{1}{\pi}\sqrt{k}\rho - \frac{1}{6\pi^2}k^{3/2}\rho^3 a_1 a_2 \right]. \quad (8.52)
$$

The particular extremum for $a_1 = a_2 \equiv a$ is a minimum of V provided that $\sqrt{k}\rho^3 > 12\pi$ (it is a maximum for $\sqrt{k}\rho^3 < 12\pi$). Consistency with the series expansion implies now

$$\sqrt{k}\rho > \frac{12\pi}{\rho^2} \gg 1. \tag{8.53}$$

This can be obtained typically for values of $\rho \simeq 3$, $k \simeq 9$, $2\pi c \simeq 9$—but, as in the former case, the range of allowed values is much wider.

Chapter 9
Eleventh Application: Topological Symmetry Breaking in Self-Interacting Theories

We consider in this chapter a self-interacting ϕ^4-theory of a massive or massless scalar field on the spacetime $\mathbb{T}^N \times \mathbb{R}^n$, $n, N \in \mathbb{N}_0$, where the torus $\mathbb{T}^N$ is assumed to have arbitrary compactification lengths. The nonrenormalized effective potential will be calculated and its precise dependence on the compactification lengths and on the mass of the field will be shown [43]. In order to determine the renormalized topologically generated mass we will restrict our considerations to $n + N = 4$ dimensions. For nonvanishing real mass of the field no symmetry breaking will occur. In contrast, for the massless scalar field one obtains that for $n = 1$ and $n = 0$ in some range of the compactification lengths symmetry breaking is possible. A numerical analysis of the topologically generated mass will be carried out.

9.1 General Considerations

Quantum field theory in partially compactified spacetime plays a fundamental role in various contexts. Let us just mention the following:

1. Finite temperature quantum field theory in the Euclidean formulation, where the imaginary time is compactified to a circle of size β, where β is the inverse temperature (see e.g. [276–280]).
2. Casimir-energy calculations, where the sign of the energy strongly depends on the number of compactified dimensions (see e.g. [10, 11, 13, 14, 42, 128–130, 148] and [281–284]).
3. Topological symmetry breaking or restoration and topological mass generation (see e.g. [285–287], and references therein).

In this chapter we will concentrate on the third point, and namely in the context of a self-interacting scalar field theory on the spacetime $\mathbb{T}^N \times \mathbb{R}^n$, $n, N \in \mathbb{N}_0$. Apart from the well known influence of topology (imposed for example by finite temperature or compactified spatial sections) on the effective mass of the field [277, 286] and [288–298]—and consequently also on particle creation [299, 300]—let us men-

E. Elizalde, *Ten Physical Applications of Spectral Zeta Functions*,
Lecture Notes in Physics 855,
DOI 10.1007/978-3-642-29405-1_9, © Springer-Verlag Berlin Heidelberg 2012

tion as a motivation for such considerations the fact that possibly the universe as a whole exhibits nontrivial topology (see [301–303] and references therein).

Most of the works mentioned above have been concerned with only one compact dimension, representing imaginary time or a compact spatial dimension. However, for example in the context of a cosmological set up, it is necessary to compactify more than one spatial dimension, as has been emphasized by Goncharov [295], who has treated the $\lambda\phi^4$-theory of a massless scalar field in static flat Clifford–Klein–Robertson–Walker spacetimes of the type $\mathbb{T}^n \times \mathbb{R}^{4-n}$, $n = 2, 3$, restricted to the case of an equilateral torus $\mathbb{T}^n$. Although in a cosmological set up the inclusion of curvature is of course necessary, in addition, this spacetime allows for high explicitness in the analysis of the effective potential and it still reveals quite interesting features. As already pointed out by Actor [286], the generalization to spacetimes of the form $\mathbb{T}^N \times \mathbb{R}^n$ with arbitrary dimensions N and n is of interest in order to analyze the dependence of the occurrence of a symmetry breaking mechanism on the dimension of spacetime.

These considerations on self-interacting $\lambda\phi^4$-theories on spacetimes $\mathbb{T}^N \times \mathbb{R}^n$ can be extended in different respects. The main emphasis here will be symmetry breaking, therefore we will concentrate on non-negative values of the square of the mass of the field. In Sect. 9.1 we will calculate the one-loop effective potential of the theory for a torus with arbitrary compactification lengths using the zeta-function regularization scheme. Technical similarities exist between [286] and the considerations here, however the details differ. In Sect. 9.2 we present a simple method to extract the topologically generated mass of the field, even without an explicit evaluation of the effective potential. A slightly different treatment for the massive and for the massless theory is necessary. After that we will perform the renormalization of the theory (Sect. 9.3). To ensure renormalizability we will restrict ourselves to $n + N = 4$ dimensions. Due to the different behavior of the massive and of the massless theory, every case is treated in its own subsection. Finally, in Sect. 9.4 we present a numerical analysis of the dependencies of the topologically generated mass on the number of compactified dimension and on the corresponding compactification lengths. For some cases, it is seen that when the compactification lengths exceed some critical value, the theory exhibits a symmetry breaking mechanism. Thanks to our explicit formulas, obtained using zeta-function regularization, the numerical values of the generated mass can be obtained with great precision and little (or no) role will be played by the specific numerical approximation method. A direct evaluation using any computation package (e.g. Mathematica or MAPLE) is straightforward.

9.2 The One-Loop Effective Potential for the Self-Interacting Theory

In this section we shall concentrate on the evaluation of the one loop effective potential in terms of Epstein-like zeta functions. The starting point of the theory is the

Lagrangian

$$\mathcal{L} = \frac{1}{2}(\partial_\mu \varphi)(\partial^\mu \varphi) - \frac{1}{2}m^2\varphi^2 - \frac{\lambda}{4!}\varphi^4, \tag{9.1}$$

with the classical potential

$$V_0[\varphi] = \frac{1}{2}m^2\varphi^2 + \frac{\lambda}{4!}\varphi^4. \tag{9.2}$$

We will consider the spacetime $\mathbb{T}^N \times \mathbb{T}^n$ with compactification lengths $L_1, \ldots, L_N, L$, and will take the limit $L \to \infty$ afterwards. In this spacetime one may assume as given a constant classical background field $\hat{\varphi}$, and the quantum fluctuations $\phi = \varphi - \hat{\varphi}$ around this background field satisfy an equation of the form

$$\left(-\Delta + M^2\right)\phi(x) = 0, \tag{9.3}$$

with an effective mass M defined by

$$M^2 = m^2 + \frac{1}{2}\lambda\hat{\varphi}^2. \tag{9.4}$$

The effective potential including one-loop quantum effects is then given by the function

$$V_{eff}(\hat{\varphi}; L_i) = \frac{1}{2}m^2\hat{\varphi}^2 + \frac{\lambda}{4!}\hat{\varphi}^4 + V\left(M^2\right), \tag{9.5}$$

where the quantum potential

$$V_n V_N V\left(M^2\right) = \frac{1}{2}\ln\det\left(\frac{-\Delta + M^2}{\mu^2}\right) \tag{9.6}$$

is the functional determinant arising from the integration over the quantum fluctuations, being V_n, V_N the volumes of the corresponding tori.

The functional determinant will be calculated using the zeta-function prescription (see Chap. 1). In this regularization scheme, it is defined by (here the limit $L \to \infty$ has been performed already)

$$V_N V\left(M^2\right) = \frac{1}{2}\left[\zeta(0; L_i)\ln\mu^2 - \zeta'(0; L_i)\right], \tag{9.7}$$

where μ is a scaling length and the prime denotes differentiation with respect to the first argument (namely, with respect to s, see (9.8)). $\zeta(s; L_i)$ is the zeta function associated with the operator (9.3), with periodic boundary conditions for the field, i.e. $\phi(x_i) = \phi(x_i + L_i)$. This means, for $\mathrm{Re}\, s > \frac{n+N}{2}$,

$$\zeta(s; L_i) = (2\pi)^{-n} \sum_{l_1,\ldots,l_N=-\infty}^{\infty} \int d^n k \cdot \left[\left(\frac{2\pi l_1}{L_1}\right)^2 + \cdots\right.$$

$$+ \left(\frac{2\pi l_N}{L_N} \right)^2 + \vec{k}^2 + M^2 \right]^{-s} \tag{9.8}$$

or, performing the $\vec{k}$-integration,

$$\zeta(s; w_i) = \left(\frac{\sqrt{\pi}}{L_1} \right)^n \frac{\Gamma(s - \frac{n}{2})}{\Gamma(s)} \left(\frac{L_1}{2\pi} \right)^{2s} Z_N^{v^2} \left(s - \frac{n}{2}; w_1, \ldots, w_N \right), \tag{9.9}$$

where we have introduced the dimensionless parameters $v^2 = c^2 + \psi^2$, with $c^2 = (\frac{L_1 m}{2\pi})^2$, $\psi^2 = \frac{\lambda}{2}(\frac{L_1 \hat{\phi}}{2\pi})^2$, furthermore $w_i = (\frac{L_1}{L_i})^2$ and the generalized Epstein zeta function

$$Z_N^{v^2}(v; w_1, \ldots, w_N) = \sum_{l_1, \ldots, l_N = -\infty}^{\infty} \left[w_1 l_1^2 + \cdots + w_N l_N^2 + v^2 \right]^{-v}, \tag{9.10}$$

valid for $\mathrm{Re}\, v > \frac{N}{2}$.

To find the effective potential (9.5) we need the derivative of (9.9) at $s = 0$. Instead of using now analytical continuations of the Epstein-like zeta-function, we will as a sort of organization of the calculation, determine first the quantum potential in terms of the properties of the function $Z_N^{v^2}$.

Using regularization techniques for the Mellin transforms (see Chap. 3 and [10, 11, 13, 14, 60, 304]), it is easy to show, that for N even the poles of order one of $Z_N^{v^2}(s; w_1, \ldots, w_N)$ are located at $s = \frac{N}{2}; \frac{N}{2} - 1; \ldots; 1$, whereas for N odd one finds $s = \frac{N}{2}; \frac{N}{2} - 1; \ldots; \frac{1}{2}; -\frac{2l+1}{2}, l \in \mathbb{N}_0$. The residuum is determined to be

$$\mathrm{Res}\, Z_N^{v^2}(j; w_1, \ldots, w_N) = \frac{(-1)^{\frac{N}{2}+j} \pi^{\frac{N}{2}}}{\sqrt{w_1 \cdots w_N}} \frac{v^{N-2j}}{\Gamma(j)(\frac{N}{2} - j)!}. \tag{9.11}$$

In addition, for $p \in \mathbb{N}_0$ one has

$$Z_N^{v^2}(-p; w_1, \ldots, w_N) = \begin{cases} 0, & \text{for } N \text{ odd,} \\ \frac{(-1)^{\frac{N}{2}} p! \pi^{\frac{N}{2}}}{\sqrt{w_1 \cdots w_N}} \frac{v^{N+2p}}{(\frac{N}{2}+p)!}, & \text{for } N \text{ even.} \end{cases} \tag{9.12}$$

Due to the different pole structure for N even and N odd, furthermore because of the different behavior of $\Gamma(s - \frac{n}{2})/\Gamma(s)$ at $s = 0$ for n even and n odd, one has to consider four different situations. Introducing PP $Z_N^{v^2}$ for the finite part of $Z_N^{v^2}$, the different results for the effective potential read:

1. For $n = 2k$, $k \in \mathbb{N}_0$, N even,

$$V_N V(M^2) = -\frac{1}{2} \left(\frac{\sqrt{\pi}}{L_1} \right)^n \frac{(-1)^k}{k!} \left\{ Z_N'^{v^2}(-k; w_1, \ldots, w_N) \right.$$

$$\left. + Z_N^{v^2}(-k; w_1, \ldots, w_N) \left[2\ln\left(\frac{L_1}{2\pi \mu} \right) + \gamma + \psi(k+1) \right] \right\}. \tag{9.13}$$

2. For $n = 2k$, $k \in \mathbb{N}_0$, N odd,

$$V_N V(M^2) = -\frac{1}{2}\left(\frac{\sqrt{\pi}}{L_1}\right)^n \frac{(-1)^k}{k!} Z_N'^{\,v^2}(-k; w_1, \ldots, w_N). \qquad (9.14)$$

3. For $n = 2k + 1$, $k \in \mathbb{N}_0$, N even,

$$V_N V(M^2) = -\frac{1}{2}\left(\frac{\sqrt{\pi}}{L_1}\right)^n \Gamma\left(-k - \frac{1}{2}\right) Z_N^{v^2}\left(-k - \frac{1}{2}; w_1, \ldots, w_N\right). \qquad (9.15)$$

4. And, for $n = 2k + 1$, $k \in \mathbb{N}_0$, N odd,

$$V_N V(M^2)$$

$$= -\frac{1}{2}\left(\frac{\sqrt{\pi}}{L_1}\right)^n \Gamma\left(-k - \frac{1}{2}\right)$$

$$\cdot \left\{ \mathrm{PP}\, Z_N^{v^2}\left(-k - \frac{1}{2}; w_1, \ldots, w_N\right) \right.$$

$$\left. + \mathrm{Res}\, Z_N^{v^2}\left(-k - \frac{1}{2}; w_1, \ldots, w_N\right)\left[2\ln\left(\frac{L_1}{2\pi\mu}\right) + \gamma + \psi\left(k + \frac{3}{2}\right)\right]\right\},$$

$$(9.16)$$

where $\psi(z) = \Gamma'(z)/\Gamma(z)$ and $\gamma = -\psi(1)$.

In order to obtain the effective potential, the remaining task is to construct analytical continuations of $Z_N^{v^2}(v; w_1, \ldots, w_N)$, (9.10), to $\mathrm{Re}\, v < \frac{N}{2}$ and to determine the properties of $Z_N^{v^2}(v; w_1, \ldots, w_N)$ needed in (9.13)–(9.16).

We now give the explicit analytical continuations of the Epstein-like zeta-functions $Z_N^{c^2}(s; w_1, \ldots, w_N)$. The first representation is useful for large values, the second one for small values of the inhomogeneity term. An analytic continuation in terms of Bessel functions is obtained in this case (the quadratic one) by making use of the Jacobi's relation between theta functions (see Chap. 2). One finds

$$Z_N^{c^2}(s; w_1, \ldots, w_N)$$

$$= \frac{\pi^{\frac{N}{2}}}{\sqrt{w_1 \cdots w_N}} \frac{\Gamma(s - \frac{N}{2})}{\Gamma(s)} c^{N-2s}$$

$$+ \frac{\pi^s}{\sqrt{w_1 \cdots w_N}} \frac{2}{\Gamma(s)} \sum_{l_1, \ldots, l_N = -\infty}^{\infty} {}' \; c^{\frac{N}{2}-s}\left[\frac{l_1^2}{w_1} + \cdots + \frac{l_N^2}{w_N}\right]^{\frac{1}{2}(s-\frac{N}{2})}$$

$$\cdot K_{\frac{N}{2}-s}\left(2\pi c\left[\frac{l_1^2}{w_1} + \cdots + \frac{l_N^2}{w_N}\right]^{\frac{1}{2}}\right), \qquad (9.17)$$

where the prime means omission of the summation index $l_1 = \cdots = l_N = 0$. Specialized to the present situation, those are just the formulas obtained in Chap. 2

already. Because of the exponential decay of the modified Bessel functions (also called Kelvin functions), this representation is obviously very valuable for a numerical analysis for large values of c. The information relevant for the topologically generated mass, (9.22)–(9.25) to follow, is relatively easy to obtain, but the expressions are rather long and will not be reproduced here.

For small values of c it is more suitable to proceed with the binomial expansion [109, 126]. One then finds the analytical continuation

$$Z_N^{c^2}(s; w_1, \ldots, w_N) = c^{-2s} + \sum_{j=0}^{\infty} (-1)^j \frac{\Gamma(s+j)}{j!\Gamma(s)} Z_N(s+j; w_1, \ldots, w_N) c^{2j}$$

(9.18)

where we have defined

$$Z_N(v; w_1, \ldots, w_N) = \sum_{l_1, \ldots, l_N = -\infty}^{\infty} {}' \left(w_1 l_1^2 + \cdots + w_N l_N^2\right)^{-v}.$$

Once more, also using this representation the properties needed in (9.21)–(9.25) can be determined and the topologically generated mass can be calculated. However, to keep the exposition clear, we will not go into the details of the calculation of the effective potential. Instead, we shall concentrate on the most physically interesting quantity, namely the mass of the classical background field $\hat{\varphi}$ generated by the quantum fluctuations.

9.3 The One-Loop Topological Mass

The one-loop quantum corrections m_T^2 to the topological mass of the background field $\hat{\varphi}$ are defined by

$$V(M^2) = \Lambda + \frac{1}{2} m_T^2 \hat{\varphi}^2 + \mathcal{O}(\hat{\varphi}^4).$$

(9.19)

To obtain the expansion (9.19) let us consider the corresponding expansion of the zeta-function $\zeta(s; L_i)$. For $m = 0$, that is also $c = 0$, the summation index $l_1 = \cdots = l_N = 0$ in (9.10) plays a crucial role. Therefore the cases $m > 0$ and $m = 0$ will be treated separately.

Let us start with $m > 0$. Using the representation of (9.9) as a Mellin-transformation, the expansion in powers of the field is easily found to be

$$\zeta(s; L_i) = \left(\frac{\sqrt{\pi}}{L_1}\right)^n \left(\frac{L_1}{2\pi}\right)^{2s} \frac{1}{\Gamma(s)}$$

$$\cdot \sum_{l_1, \ldots, l_N = -\infty}^{\infty} \int_0^{\infty} dt \, t^{s-\frac{n}{2}-1} \exp\left[-\left(w_1 l_1^2 + \cdots + w_N l_N^2 + c^2\right)t\right]$$

$$\cdot \left[1 - \psi^2 t + \mathcal{O}(\hat{\varphi}^4)\right]$$

$$= \left(\frac{\sqrt{\pi}}{L_1}\right)^n \left(\frac{L_1}{2\pi}\right)^{2s} \frac{1}{\Gamma(s)} \left[\Gamma\left(s - \frac{n}{2}\right) Z_N^{c^2}\left(s - \frac{n}{2}; w_1, \ldots, w_N\right)\right.$$

$$\left. - \frac{\lambda}{2}\left(\frac{L_1}{2\pi}\right)^2 \Gamma\left(s - \frac{n}{2} + 1\right) Z_N^{c^2}\left(s - \frac{n}{2} + 1; w_1, \ldots, w_N\right) \hat{\varphi}^2 + \mathcal{O}(\hat{\varphi}^4)\right].$$

$$(9.20)$$

Using (9.20) it is easy to obtain the unrenormalized effective potential up to $\mathcal{O}(\hat{\varphi}^4)$. Here we will give only the topological mass m_T^2, which in terms of properties of the Epstein-like zeta-functions for the different cases reads:

1. Case $n = 0$

$$V_N m_T^2 = \frac{\lambda}{2}\left(\frac{L_1}{2\pi}\right)^2 \left[\mathrm{PP}\, Z_N^{c^2}(1; w_1, \ldots, w_N)\right.$$

$$\left. + \mathrm{Res}\, Z_N^{c^2}(1; w_1, \ldots, w_N) \ln\left(\frac{L_1}{2\pi\mu}\right)^2\right]. \qquad (9.21)$$

2. Case $n = 2k$, $k \in \mathbb{N}$, N even,

$$V_N m_T^2 = -\frac{\lambda}{2} k \left(\frac{L_1}{2\pi}\right)^2 \left(\frac{\sqrt{\pi}}{L_1}\right)^n \frac{(-1)^k}{k!} \left\{Z_N'^{c^2}(-k+1; w_1, \ldots, w_N)\right.$$

$$\left. + Z_N^{c^2}(-k+1; w_1, \ldots, w_N) \left[2\ln\left(\frac{L_1}{2\pi\mu}\right) + \gamma + \psi(k)\right]\right\}. \quad (9.22)$$

3. Case $n = 2k$, $k \in \mathbb{N}$, N odd,

$$V_N m_T^2 = -\frac{\lambda}{2} k \left(\frac{L_1}{2\pi}\right)^2 \left(\frac{\sqrt{\pi}}{L_1}\right)^n \frac{(-1)^k}{k!} Z_N'^{c^2}(-k+1; w_1, \ldots, w_N). \quad (9.23)$$

4. Case $n = 2k + 1$, $k \in \mathbb{N}_0$, N even,

$$V_N m_T^2 = \frac{\lambda}{2}\left(\frac{L_1}{2\pi}\right)^2 \left(\frac{\sqrt{\pi}}{L_1}\right)^n \Gamma\left(-k + \frac{1}{2}\right) Z_N^{c^2}\left(-k + \frac{1}{2}; w_1, \ldots, w_N\right). \quad (9.24)$$

5. Case $n = 2k + 1$, $k \in \mathbb{N}_0$, N odd,

$$V_N m_T^2 = \frac{\lambda}{2}\left(\frac{L_1}{2\pi}\right)^2 \left(\frac{\sqrt{\pi}}{L_1}\right)^n \Gamma\left(-k + \frac{1}{2}\right) \left\{\mathrm{PP}\, Z_N^{c^2}\left(-k + \frac{1}{2}; w_1, \ldots, w_N\right)\right.$$

$$\left. + \mathrm{Res}\, Z_N^{v^2}\left(-k + \frac{1}{2}; w_1, \ldots, w_N\right) \left[2\ln\left(\frac{L_1}{2\pi\mu}\right) + \gamma + \psi\left(k + \frac{1}{2}\right)\right]\right\}.$$

$$(9.25)$$

Using the results above, analytical expressions for the generated mass m_T^2 are now easy to find. But the results are rather long and not especially illuminating, so we will not write them down here. However, after going through the renormalization procedure in $n + N = 4$ dimensions the analytical representation of the Epstein-like zeta-functions will be used explicitly, because so it will be possible to prove that m_T^2 in (9.21)–(9.25) is always positive.

Let us now consider the massless case $m = 0$. For that we introduce

$$Z_N(s; w_1, \dots, w_N) = \sum_{l_1,\dots,l_N=-\infty}^{\infty}{}' (w_1 l_1^2 + \cdots + w_N l_N^2)^{-s} \qquad (9.26)$$

where the prime means omission of the zero mode. Then expansion (9.20) has to be replaced with

$$\zeta(s; L_i) = \left(\frac{\sqrt{\pi}}{L_1}\right)^n \left(\frac{L_1}{2\pi}\right)^{2s} \frac{\Gamma(s - \frac{n}{2})}{\Gamma(s)} v^{n-2s} + \left(\frac{\sqrt{\pi}}{L_1}\right)^n \left(\frac{L_1}{2\pi}\right)^{2s} \frac{1}{\Gamma(s)}$$

$$\cdot \left[\Gamma\left(s - \frac{n}{2}\right) Z_N\left(s - \frac{n}{2}; w_1, \dots, w_N\right) \right.$$

$$\left. - \frac{\lambda}{2}\left(\frac{L_1}{2\pi}\right)^2 \Gamma\left(s - \frac{n}{2} + 1\right) Z_N\left(s - \frac{n}{2} + 1; w_1, \dots, w_N\right) \hat{\varphi}^2 + \mathcal{O}(\hat{\varphi}^4) \right].$$

$$(9.27)$$

where the first term reveals the special role played by the summation index $l_1 = \cdots = l_N = 0$ in (9.10). For $n \neq 2$ this term does not contribute to the topologically generated mass and with the replacement $Z_N^{c^2} \rightarrow Z_N$ (9.21)–(9.25) remain true. However, for the case $n = 2$ the mass m_T^2 cannot be calculated, as a result of the well-known infrared problems in two dimensions.

9.4 Renormalization of the Theory

Up to now the analysis has been performed for general dimensions n and N. Now, in order to implement the renormalization procedure we will restrict ourselves to $n + N = 4$ dimensions. As is seen in (9.13) and (9.16) (resp. (9.21), (9.22) and (9.25)) the effective potential (resp. the topologically generated mass m_T^2) depends on the arbitrary scaling length μ. However, with the help of (9.11) and (9.12), one can show, that the prefactors of the terms depending on the scale μ do not depend on the compactification lengths of the torus. So it is enough to perform the renormalization of the Euclidean spacetime (for similar topologies this has already been observed for example by Toms [290, 305]). Let us first consider the case $m > 0$. Then more specifically, this means that we can define the renormalized effective potential as

$$V_{eff}^{(ren)}(\hat{\varphi}, L_i) = \delta C + \frac{1}{2}\left(m^2 + \delta m^2\right)\hat{\varphi}^2 + \frac{1}{4!}(\lambda + \delta\lambda)\hat{\varphi}^4 + V(M^2), \qquad (9.28)$$

where the counterterms are fixed by the renormalization conditions

$$0 = V_{\text{eff}}^{(ren)}(\hat{\varphi} = 0, L_i \to \infty), \tag{9.29}$$

$$m^2 = \left. \frac{\partial^2 V_{\text{eff}}^{(ren)}(\hat{\varphi}, L_i)}{\partial \hat{\varphi}^2} \right|_{\hat{\varphi}=0, L_i \to \infty}, \tag{9.30}$$

$$\lambda = \left. \frac{\partial^4 V_{\text{eff}}^{(ren)}(\hat{\varphi}, L_i)}{\partial \hat{\varphi}^4} \right|_{\hat{\varphi}=\hat{\varphi}_1, L_i \to \infty}. \tag{9.31}$$

We find

$$\delta C = \frac{m^4}{64\pi^2} \left[\frac{3}{2} - \ln(m^2\mu^2) \right], \tag{9.32}$$

$$\delta m^2 = \frac{\lambda m^2}{32\pi^2} \left[1 - \ln(m^2\mu^2) \right], \tag{9.33}$$

$$\delta \lambda = \frac{\lambda^2}{32\pi^2} \left[\frac{\lambda^2 \hat{\varphi}_1^4}{M_1^4} - \frac{6\lambda \hat{\varphi}_1^2}{M_1^2} - 3\ln(M_1^2\mu^2) \right], \tag{9.34}$$

with $M_1^2 = m^2 + \frac{1}{2}\lambda\hat{\varphi}_1^2$.

Thus, the renormalized effective potential reads

$$V_{\text{eff}}^{(ren)}(\hat{\varphi}; L_i) = \frac{1}{2}m^2\hat{\varphi}^2 + \frac{1}{4!}\lambda\hat{\varphi}^4 - \frac{1}{64\pi^2}\left\{ M^4\left[\ln\left(\frac{L_1 m}{2\pi}\right)^2 + \gamma + \psi\left(\frac{n}{2}+1\right) \right] \right.$$

$$+ \frac{1}{4}\lambda^2\hat{\varphi}^4 \ln\left(\frac{M_1}{2\pi m}\right)^2 - \frac{3}{2}m^4 - \lambda m^2\hat{\varphi}^2 - \frac{\lambda^2\hat{\varphi}^4}{12}\lambda\hat{\varphi}_1^2 \left[\frac{\lambda\hat{\varphi}_1^2}{M_1^4} - \frac{6}{M_1^2}\right] \right\}$$

$$- \frac{1}{2V_N}\left(\frac{\sqrt{\pi}}{L_1}\right)^n$$

$$\cdot \begin{cases} \frac{(-1)^{\frac{n}{2}}}{(\frac{n}{2})!} Z_N'^{v^2}(-\frac{n}{2}; w_1, \ldots, w_N), & n \text{ even,} \\ \Gamma(-\frac{n}{2}) \, PP \, Z_N^{v^2}(-\frac{n}{2}; w_1, \ldots, w_N), & n \text{ odd.} \end{cases} \tag{9.35}$$

For the renormalized topologically generated mass, we obtain

$$m_{T,ren}^2 = m^2 - \frac{\lambda m^2}{32\pi^2}\left[\ln\left(\frac{L_1 m}{2\pi}\right)^2 - 1 + (\delta_{n,0} - 1)\left\{ \gamma + \psi\left(\frac{n}{2}\right) \right\} \right]$$

$$+ \frac{\lambda}{2V_N}\left(\frac{L_1}{2\pi}\right)^2\left(\frac{\sqrt{\pi}}{L_1}\right)^n$$

$$\cdot \begin{cases} PP\ Z_4^{c^2}(1; w_1, \ldots, w_4), & n = 0, \\ Z_2'^{c^2}(0; w_1, w_2), & n = N = 2, \\ \Gamma(1 - \tfrac{n}{2})\ PP\ Z_N^{c^2}(1 - \tfrac{n}{2}; w_1, \ldots, w_N), & n \text{ odd}. \end{cases} \quad (9.36)$$

At this point, let us use the analytical continuation of $Z_N^{c^2}$, (9.17). Introducing the dimensionless quantities

$$x \equiv \left(\frac{L_1 m}{4\pi}\right), \qquad y \equiv \frac{\lambda}{16\pi^2}, \qquad z \equiv \left(\frac{L_1 m_{T,ren}}{4\pi}\right)^2, \quad (9.37)$$

the topologically generated mass for all cases reads

$$z = x + \frac{y\sqrt{x}}{2\pi} \sum_{n_1,\ldots,n_N=-\infty}^{\infty}{}' \left[n_1^2/w_1 + \cdots + n_N^2/w_N\right]^{-\frac{1}{2}}$$
$$\cdot K_1\left(4\pi\sqrt{x}\left[n_1^2/w_1 + \cdots + n_N^2/w_N\right]^{\frac{1}{2}}\right). \quad (9.38)$$

It is seen, that for $x > 0$ one concludes $z > 0$. This means, that for positive square of the mass of the classical field the quantum fluctuations do not lead to a symmetry breaking mechanism. This changes if we consider a massless scalar field as will be shown in the following section.

9.5 Symmetry Breaking Mechanism for a Massless Scalar Field

Without going once more into the details of the calculation, we will just state the final results for the topologically generated mass in the case $m = 0$. We find

1. Case $n = 3$, $N = 1$:

$$m_T^2 = -\frac{\lambda}{4L_1^2} Z_1\left(-\frac{1}{2}; 1\right) = \frac{\lambda}{24L_1^2}. \quad (9.39)$$

2. Case $n = 1$, $N = 3$:

$$m_T^2 = \frac{\lambda}{8\pi L_2 L_3} Z_3\left(\frac{1}{2}; w_1, w_2, w_3\right). \quad (9.40)$$

3. Case $n = 0$, $N = 4$:

$$m_T^2 = \frac{\lambda L_1}{8\pi^2 L_2 L_3 L_4} Z_4(1; w_1, w_2, w_3, w_4). \quad (9.41)$$

The first of these equations (9.39) is just the well known finite temperature result given for example in [277, 293, 294], which is certainly positive and therefore of lesser interest for our present considerations. However, from (9.40) and (9.41) we see that the sign depends on the compactification lengths. This is an interesting result [43]. For example the equilateral torus leads in both cases to a negative m_T^2 (the case $N = 3$ has also been treated in [295] and the results here are in numerical agreement with that reference). Varying the compactification lengths also symmetry restoration is possible [43].

Summarizing the results of the calculation, in the space of compactification lengths a critical region can be characterized which encloses the value $L_1 = \cdots = L_N$ and a clear transition is observed from negative (region around $L_1 = \cdots = L_N$) to positive values of m_T^2. This can be easily checked, e.g. for $n = 1, N = 3$ and $n = 0, N = 4$. In all cases, the region with ($m_T^2 \leq 0$) is finite and symmetric around the point with equal L_is [43].

The numerical analysis above is based on the expressions for Z_3 and Z_4 that we now write explicitly, where an analytical continuation for Z_N has been used in order to derive analytical expressions for m_T^2. The analytical continuation of the Epstein-like zeta-function $Z_N(s; w_1, \ldots, w_N)$ and its properties needed in (9.40) and (9.41) to give the numerical analysis of the topologically generated mass m_T^2.

The analytical continuation of (9.26) is (see Chap. 2)

$$
Z_N(s; w_1, \ldots, w_N)
$$

$$
= \frac{2}{w_N^s} \zeta_R(2s) + \sqrt{\frac{\pi}{w_N}} \frac{\Gamma(s - \frac{1}{2})}{\Gamma(s)} Z_{N-1}\left(s - \frac{1}{2}; w_1, \ldots, w_{N-1}\right)
$$

$$
+ \frac{4\pi^s}{\Gamma(s)\sqrt{w_N}} \sum_{n=1}^{\infty} \sum_{n_1,\ldots,n_{N-1}=-\infty}^{\infty} {}' \left(\frac{[w_1 n_1^2 + \cdots + w_{N-1} n_{N-1}^2] w_N}{n^2}\right)^{\frac{1}{2}(\frac{1}{2}-s)}
$$

$$
\cdot K_{\frac{1}{2}-s}\left(\frac{2\pi n}{\sqrt{w_N}} \sqrt{w_1 n_1^2 + \cdots + w_{N-1} n_{N-1}^2}\right). \tag{9.42}
$$

Using (9.42) the relevant results for the topologically generated mass read

$$
Z_3\left(\frac{1}{2}; 1, w_2, w_3\right) = \frac{1}{\sqrt{w_3}}\left[2\gamma + \ln\left(\frac{w_2}{w_3}\right) - 2\ln(4\pi) + \frac{\pi}{3\sqrt{w_2}}\right]
$$

$$
+ \frac{8}{\sqrt{w_2 w_3}} \sum_{n,n_1=1}^{\infty} \left(\frac{n_1\sqrt{w_2}}{n}\right)^{\frac{1}{2}} K_{\frac{1}{2}}\left(\frac{2\pi n n_1}{\sqrt{w_2}}\right)
$$

$$
+ \frac{4}{\sqrt{w_3}} \sum_{n=1}^{\infty} \sum_{n_1,n_2=-\infty}^{\infty} {}' K_0\left(\frac{2\pi n}{\sqrt{w_3}} \sqrt{n_1^2 + n_2^2 w_2}\right) \tag{9.43}
$$

and

$$
Z_4(1; 1, w_2, w_3, w_4)
$$

$$
= \frac{\pi}{\sqrt{w_4}}\left[\frac{2\gamma}{\sqrt{w_3}} + \frac{2}{\sqrt{w_3}} \ln\left(\frac{\sqrt{w_2}}{4\pi\sqrt{w_3}}\right) + \frac{\pi}{3\sqrt{w_2 w_3}} + \frac{\pi}{3\sqrt{w_4}}\right]
$$

$$+ \frac{8\pi}{\sqrt{w_2 w_3 w_4}} \sum_{n,n_1=1}^{\infty} \left(\frac{n_1 \sqrt{w_2}}{n} \right)^{\frac{1}{2}} K_{\frac{1}{2}} \left(\frac{2\pi n n_1}{\sqrt{w_2}} \right)$$

$$+ \frac{4\pi}{\sqrt{w_3 w_4}} \sum_{n=1}^{\infty} \sum_{n_1,n_2=-\infty}^{\infty} {}' K_0 \left(\frac{2\pi n}{\sqrt{w_3}} \sqrt{n_1^2 + n_2^2 w_2} \right)$$

$$+ \frac{4\pi}{\sqrt{w_4}} \sum_{n=1}^{\infty} \sum_{n_1,n_2,n_3=-\infty}^{\infty} {}' \left(\frac{[n_1^2 + w_2 n_2^2 + w_3 n_3^2] w_4}{n^2} \right)^{-\frac{1}{4}}$$

$$\cdot K_{\frac{1}{2}} \left(\frac{2\pi n}{\sqrt{w_4}} \sqrt{n_1^2 + w_2 n_2^2 + w_3 n_3^2 + w_4 n_4^2} \right), \qquad (9.44)$$

which are very convenient expressions that can be evaluated numerically (see [43] for details of this calculation).

To summarize, in this chapter we have addressed the issue of topological mass generation for the case of a self-interacting scalar field theory on spacetimes of the type $\mathbb{T}^N \times \mathbb{R}^n$, $n, N \in \mathbb{N}_0$. We have been concerned with symmetry breaking questions, therefore real masses for the fields have been taken into account only. After showing in some detail the calculation of the effective potential, we have analyzed the behavior of the basic, physically interesting quantity, namely the mass of the classical background field generated by the quantum fluctuations. A detailed derivation of the renormalized mass has been carried out. It only involves zeta-function techniques and the final results are expressed as convergent series which can be added up numerically to any desired approximation. For the case $m^2 > 0$ (9.38) represents the central result. It shows that for $m^2 > 0$ symmetry breaking is not possible, irrespective of the values of the compactification lengths L_i and of the parameter m^2. One is able to establish this remarkable fact [43] in a rigorous way thanks again to the quick convergence of the non-trivial expansions for the zeta-functions.

The corresponding results for $m^2 = 0$ are given by (9.39)–(9.41). They are even more interesting than those for the massive case. Here the occurrence of symmetry breaking for $n = 1$ and $n = 0$ is clearly established and, moreover, it is explicitly shown to depend on the values of the compactification lengths L_i [43]. Furthermore, after symmetry breaking has occurred, continuing along the same path in the parameter space of compactification lengths one is then led to symmetry restoration in a natural way. There is a clear reason for this fact, as soon as one realizes the symmetric role played by the different compactification lengths of the multidimensional torus.

Chapter 10
Twelfth Application: Cosmology and the Quantum Vacuum

Zeta regularization has proven to be a powerful and reliable tool for the regularization of the vacuum energy density in ideal situations. With the Hadamard complement, it has been shown to provide finite (and meaningful) answers too in more involved cases, as when imposing physical boundary conditions (BCs) in two- and higher-dimensional surfaces (being able to mimic in a very convenient way other *ad hoc* cut-offs, as non-zero depths), as we have seen before (Chap. 7).

Also these techniques have been used in calculations of the contribution of the vacuum energy of the quantum fields pervading the universe to the cosmological constant (cc). Naive calculations of the absolute contributions of the known fields lead to a value which is off by roughly 120 orders of magnitude, as compared with observational tests, what is known as the *new cosmological constant problem*. This is difficult to solve and many authors still stick to the old problem to try to prove that basically its value is zero with some perturbations thereof leading to the (small) observed result (Burgess et al., Padmanabhan, etc.) We address this issue, in this last chapter, in a somewhat similar way, by considering the *additional* contributions to the cc that may come from the possibly non-trivial topology of space and from specific boundary conditions imposed on braneworld and other seemingly reasonable models that are being considered in the literature (mainly with other purposes too)—kind of a Casimir effect at cosmological scale. If the ground value of the cc would be indeed zero, we would then be left with this perturbative quantity coming from the topology or BCs.

We show here that this value acquires the correct order of magnitude (and has the right sign, what is also non-trivial)—corresponding to the one coming from the observed acceleration in the expansion of our universe—in a number of quite reasonable models involving small and large compactified scales and/or brane BCs, and supergravitons.

E. Elizalde, *Ten Physical Applications of Spectral Zeta Functions*,
Lecture Notes in Physics 855,
DOI 10.1007/978-3-642-29405-1_10, © Springer-Verlag Berlin Heidelberg 2012

10.1 On the Reality of the Vacuum Fluctuations

In ordinary QFT, one cannot give a meaning to the *absolute* value of the zero-point energy. Any physically measurable effect comes as an energy *difference* between two situations, such as a quantum field in curved space as compared with the same field in flat space, or one satisfying BCs on some surface as compared with the same in its absence, etc. This difference is the Casimir energy: $E_C = E_0^{BC} - E_0 = \frac{1}{2}(\text{tr } H^{BC} - \text{tr } H)$. But here a problem appears. Imposing mathematical boundary conditions (BCs) on physical quantum fields turns out to be a highly non-trivial act. This was discussed in detail in a paper by Deutsch and Candelas [213]. These authors quantized em and scalar fields in the region near an arbitrary smooth boundary, and calculated the renormalized vacuum expectation value of the stress-energy tensor, to find out that the energy density diverges as the boundary is approached. Therefore, regularization and renormalization did not seem to cure the problem with infinities in this case and an infinite *physical* energy was obtained if the mathematical BCs were to be fulfilled. However, the authors argued that surfaces have non-zero depth, and its value could be taken as a handy dimensional cutoff in order to regularize the infinities (see Sect. 7.3). Just two years after Deutsch and Candelas' work, Kurt Symanzik carried out a rigorous analysis of QFT in the presence of boundaries [214]. Prescribing the value of the quantum field on a boundary means using the Schrödinger representation, and Symanzik was able to show rigorously that such representation exists to all orders in the perturbative expansion. He showed also that the field operator being diagonalized in a smooth hypersurface differs from the usual renormalized one by a factor that diverges logarithmically when the distance to the hypersurface goes to zero. This requires a precise limiting procedure and point splitting to be applied. In any case, the issue was proven by him to be perfectly meaningful within the domains of renormalized QFT. In this case the BCs and the hypersurfaces themselves were treated at a pure mathematical level (zero depth) by using Dirac delta functions.

Recently, a new approach to the problem has been postulated [215–218]. As crudely stated by Jaffe [306], experimental confirmation of the Casimir effect does not establish by itself the reality of zero point fluctuations. He explains this via the example of the electromagnetic field, where the energy of a smooth charge distribution, $\rho(x)$, can be precisely calculated from the energy stored in the electric field, a formula which arguably cannot be taken as evidence for the electric field itself being real. Fortunately, propagating electromagnetic waves are detected all the time. The moral: in the case of the Casimir forces one should look for *direct* evidence of vacuum fluctuations. Have they been found yet? As of today, the answer is very controversial. Since GR has much wider consensus, a search at the cosmological level is proposed. In fact, almost everybody admits that any sort of energy will always gravitate [307]. Thus, the energy density of the vacuum, more precisely, the vacuum expectation value of the stress-energy tensor,

$$\langle T_{\mu\nu} \rangle \equiv -\mathcal{E} g_{\mu\nu}, \qquad (10.1)$$

will appear on the rhs of Einstein's equations

$$R_{\mu\nu} - \frac{1}{2} g_{\mu\nu} R = -8\pi G(\tilde{T}_{\mu\nu} - \mathcal{E} g_{\mu\nu}).$$ (10.2)

It therefore affects *cosmology*: there is a contribution $\tilde{T}_{\mu\nu}$ of excitations above the vacuum, equivalent to a *cosmological constant* $\lambda = 8\pi G \mathcal{E}$. Reliable data yield [308–310]

$$\lambda = \left(2.14 \pm 0.13 \times 10^{-3} \text{ eV}\right)^4 \sim 4.32 \times 10^{-9} \text{ erg/cm}^3.$$ (10.3)

At issue is then the belief that zero point fluctuations will contribute in an essential way to the cosmological constant (cc), e.g., they will be of the same order of magnitude.

Different rigorous techniques have been used in order to perform this calculation, the result being that the absolute contributions of the known quantum fields (all of which couple to gravity) lead to a value which is off by roughly 120 orders of magnitude—kind of a modern (and indeed very thick!) ether. Extremely severe cancellations should occur. Observational tests, as advanced, see nothing (or very little) of it, what leads to the so-called cosmological constant problem [311–313]. This problem is at present very difficult to solve and we will here *not* address such hard question directly. Some *almost* successful attempts at solving the problem deserve to be mentioned, as the clever approaches by Baum and Hawking, and Polchinski's phase ambiguity found in Coleman's solution [314–317].

What we do consider (see also [318])—with relative success in quite different approaches—is the *additional* contribution to the cc coming from the *non-trivial topology* of space or from specific *boundary conditions* imposed on braneworld and other models. This can be viewed as kind of a Casimir effect at cosmological scale: a *cosmo-topological Casimir effect* [319]. Assuming someone will be able to prove (some day) that the ground value of the cc is *zero* (as many had suspected until very recently),[12] we will be left with this incremental value coming from the topology or BCs. We will show that this value has the correct order of magnitude, e.g., the one coming from the observed acceleration in the expansion of our universe, in three different types of models, involving: (a) small and large compactified scales, (b) dS and AdS worldbranes, and (c) supergravitons. The case of moving boundaries seems to present quite severe difficulties, though a promising approach in order to deal with them has been issued [320].

10.2 On the Curvature and Topology of Space

The Friedmann–Robertson–Walker (FRW) model, which can be derived as the *only* family of solutions to the Einstein's equations compatible with the assumptions of

---

[12]What would, by the way, correspond to the convention of normal ordering in QFT in ordinary, Euclidean backgrounds.

homogeneity and *isotropy* of space, is the generally accepted model of the cosmos. But the FRW is a family with a free parameter, k, the curvature, that can be either positive, negative or zero (the flat or Euclidean case). This curvature, or equivalently the curvature radius, R, is not fixed by the theory and should be matched with cosmological observations. Moreover, the FRW model, and Einstein's equations themselves, can only provide local properties, not global ones, so they cannot tell about the overall topology of our world: is it closed or open? finite or infinite? Even being quite clear that it is, in any case, extremely large—and possibly the human species will never reach more than an infinitesimally tiny part of it—the question is very appealing to any (note that this discussion concerns only three-dimensional space curvature and topology, time will not be involved).

10.2.1 On the Curvature

Serious attempts to measure the possible curvature of the space we live in go back to Gauss, who measured the sum of the three angles of a big triangle with vertices on the picks of three far away mountains (Brocken, Inselberg, and Hohenhagen). He was looking for evidence that the geometry of space is non-Euclidean. The idea was brilliant, but condemned to failure: one needs a much bigger triangle to try to find the possible non-zero curvature of space. Now cosmologist have recently measured the curvature radius R by using the largest triangle available, namely one with us at one vertex and with the other two on the hot opaque surface of the ionized hydrogen that delimits our visible universe and emits the CMB radiation (some 3 to 4×10^5 years after the Big Bang) [321]. The CMB maps exhibit hot and cold spots. It can be shown that the characteristic spot angular size corresponds to the first peak of the temperature power spectrum, which is reached for an angular size of $.5°$ (approximately the one subtended by the Moon) if space is flat. If it has a positive curvature, spots should be larger (with a corresponding displacement of the position of the peak), and correspondingly smaller for negative curvature. The joint analysis of the considerable amount of data obtained by balloon experiments (BOOMERanG, MAXIMA, DASI) [322–324], combined with galaxy clustering data, has produced a lower bound for $|R| > 20h^{-1}$ Gpc, i.e. twice as large as the radius of the observable universe, of about $R_U \simeq 9h^{-1}$ Gpc.

10.2.2 On the Topology

Let us repeat that GR does not prescribe the topology of the universe, or its being finite or not, and the universe could perfectly be flat and finite. The simplest non-trivial model from the theoretical viewpoint is the toroidal topology. Traces for this and more elaborated ones, as negatively curved but compact spaces, have been profusely investigated, and some circles in the sky with near identical temperature patterns were identified [325]. And yet more papers appear from time to

time proposing a new topology [326]. However, to summarize all these efforts and
the observational situation, and once the numerical data are interpreted without bias
(what sometimes was not the case, and led to erroneous conclusions), it seems at
present that available data point towards a very large (we may call it *infinite*) flat
space.

10.3   Vacuum Energy Fluctuations and the Cosmological Constant

The issue of the cc has got renewed thrust from the observational evidence of an
acceleration in the expansion of our Universe, initially reported by two different
groups [327, 328] whose team leaders have been awarded this year's Nobel Prize
in Physics. There was some controversy on the reliability of the results obtained
from those observations and on its precise interpretation, by a number of different
reasons. Anyway, after new data has been gathered and studied using alternative, un-
correlated methods, there is now consensus among the community of cosmologists
that actually the acceleration is there, and that it does have the order of magnitude
obtained in the above mentioned observations [329–331]. As a consequence, many
theoreticians have urged to try to explain this fact, and also to try to reproduce the
precise value of the cc coming from these observations [332–334].

Now, as crudely stated by Weinberg [335], it is even more difficult to explain why
the cc is so small but non-zero, than to build theoretical models where it exactly
vanishes [336–343]. Rigorous calculations performed in quantum field theory on
the vacuum energy density, ρ_V, corresponding to quantum fluctuations of the fields
we observe in nature, lead to values that are over 120 orders of magnitude in excess
of those allowed by observations of the spacetime around us. In fact, if the cc gets
contributions from zero point fluctuations [344]

$$E_0 = \frac{\hbar c}{2} \sum_n \omega_n, \quad \omega = k^2 + m^2/\hbar^2, \ k = 2\pi/\lambda. \qquad (10.4)$$

Evaluating in a box and putting a cut-off at maximum k_{max} corresponding to reliable
QFT physics (e.g., the Planck energy)

$$\rho \sim \frac{\hbar k_{Planck}^4}{16\pi^2} \sim 10^{123} \rho_{obs}. \qquad (10.5)$$

Assuming one will be able to prove (in the future) that the ground value of the
cc is *zero* (as many suspected until recently), we will be left with this *incremental
value* coming from the topology or BCs. This sort of two-step approach to the cc is
becoming more and more popular recently as a way to try to solve this very difficult
issue [345–349]. We have seen, using different examples, that this value acquires in
fact the correct order of magnitude—corresponding to the one coming from the observed acceleration in the expansion of our universe—under some reasonable conditions. We put forward a quite simple and primitive idea (but, for the same reason,

of possibly far reaching consequences), related with the *global* topology of the universe [350] and in connection with the possibility that a faint scalar field pervading the universe could exist. Fields of this kind are ubiquitous in inflationary models, quintessence theories, and the like. In other words, we do not pretend to solve the old problem of the cc, not even to contribute significantly to its understanding, but just to present simple and usual models which show that the right order of magnitude of (some contributions to) ρ_V which lie in the precise range deduced from the astrophysical observations are not difficult to get. To say it in different words, we only address here the 'second stage' of what has been termed by Weinberg [335] the *new* cc problem.

10.4 Simple Model with Large and Small Compactified Dimensions

Consider a universe with a spacetime such as: $\mathbb{R}^{d+1} \times \mathbb{T}^p \times \mathbb{T}^q, \mathbb{R}^{d+1} \times \mathbb{T}^p \times \mathbb{S}^q, \ldots$, which are very simple models for the spacetime topology [351]. A free scalar field pervading the universe will satisfy $(-\Box + M^2)\phi = 0$, restricted by the appropriate boundary conditions (e.g., periodic, in the first case). Here, $d \geq 0$ stands for a possible number of non-compactified dimensions. Recall that the physical contribution to the vacuum or zero-point energy $\langle 0|H|0 \rangle$ (H is the Hamiltonian and $|0\rangle$ the vacuum state) is obtained after subtracting $E_C = \langle 0|H|0 \rangle|_R - \langle 0|H|0 \rangle|_{R\to\infty}$ (R being a compactification length), what gives rise to the finite value of the Casimir energy E_C, which will depend on R, after a regularization/renormalization procedure is carried out. We discuss the Casimir energy *density* $\rho_C = E_C/V$, for either a finite or an infinite volume of the spatial section of the universe.[13] In terms of the spectrum: $\langle 0|H|0 \rangle = \frac{1}{2}\sum_n \lambda_n$, the sum over n involving, in general, several continuum and several discrete indices.

The physical vacuum energy density corresponding to the contribution of a scalar field, ϕ in a (partly) compactified spatial section of the universe is[14]

$$\rho_\phi = \frac{1}{2}\sum_k \frac{1}{\mu}(k^2 + M^2)^{1/2}, \tag{10.6}$$

where μ is the usual mass-dimensional parameter to render the eigenvalues dimensionless (we take $\hbar = c = 1$ but will insert the dimensionfull units at the end). The mass M of the field will be kept different from zero (a tiny mass can never be excluded) and its allowed value will be constrained later. A lack of this simplified model: the coupling of the scalar field to gravity should be considered (see, e.g.,

---

[13]From now on we assume that all diagonalizations already correspond to energy densities, and the volume factors will be replaced at the end.

[14]Note that this is just the contribution to ρ_V coming from this field; there might be other, in general.

[352, 353] and the references therein). However, taking it into account does not change the order of magnitude of the results. The renormalization of the model is rendered much more involved, and one must enter a discussion on the orders of magnitude of the different contributions, which yields, in the end, an ordinary perturbative expansion, the coupling constant being finally re-absorbed into the mass of the scalar field. Owing, essentially, to the smallness of the resulting mass for the scalar field, one can prove that, quantitatively, the difference in the final result is of some percent only. Another consideration: our model is stationary, while the universe is in accelerated expansion. Again, in a first approach this effect will be dismissed, at the level of our order-of-magnitude calculation, since this contribution is clearly less than the one we will finally get—taken the present value of the expansion rate $\Delta R/R \sim 10^{-10}$ per year, or from direct consideration of the Hubble coefficient. In any case, these refinements are left for future work. Here, to focus just on the essential idea, we perform a static calculation and the value of the Casimir energy density and cc contribution to be obtained will correspond to the present epoch. They are certainly bound to change with time.

10.4.1 Regularization of the Vacuum Energy Density

For a (p, q)-toroidal universe, with p the number of large and q of small dimensions:

$$\rho_\phi = \frac{1}{a^p b^q} \sum_{n_p, m_q = -\infty}^{\infty} \left(\frac{1}{a^2} \sum_{j=1}^{p} n_j^2 + \frac{1}{b^2} \sum_{h=1}^{q} m_h^2 + M^2 \right)^{(d+1)/2+1}, \qquad (10.7)$$

which corresponds to all large (resp. all small) compactification scales being the same. The squared mass of the field should be divided by $4\pi^2\mu^2$, but we have renamed it again M^2 to simplify. We also dismiss the mass-dimensional factor μ, easy to recover later.

For a $(p$-toroidal, q-spherical)-universe,

$$\rho_\phi = \frac{1}{a^p b^q} \sum_{n_p = -\infty}^{\infty} \sum_{l=1}^{\infty} P_{q-1}(l) \left(\frac{4\pi^2}{a^2} \sum_{j=1}^{p} n_j^2 + \frac{l(l+q)}{b^2} + M^2 \right)^{(d+1)/2+1} \qquad (10.8)$$

$P_{q-1}(l)$ being a polynomial in l of degree $q-1$. We assume that $d = 3 - p$ is the number of non-compactified, large spatial dimensions, and ρ_ϕ needs to be regularized. We use the zeta function [10, 11, 13, 14], taking advantage of our expressions in [123, 124, 354]. No further subtraction or renormalization is needed (the subtraction at infinity is zero, and not even a finite renormalization shows up). Using the mentioned formulas, that generalize the Chowla–Selberg expression to encompass (10.7) and (10.8), we can provide arbitrarily accurate results (even for different values of the compactification radii [355, 356]).

For the first case, (10.7), we obtain

$$\rho_\phi = -\frac{1}{a^p b^{q+1}} \sum_{h=0}^{p} \binom{p}{h} 2^h \sum_{n_h=1}^{\infty} \sum_{m_q=-\infty}^{\infty} \sqrt{\frac{\sum_{k=1}^{q} m_k^2 + M^2}{\sum_{j=1}^{h} n_j^2}}$$

$$\cdot K_1 \left[\frac{2\pi a}{b} \sqrt{\sum_{j=1}^{h} n_j^2 \left(\sum_{k=1}^{q} m_k^2 + M^2 \right)} \right]. \tag{10.9}$$

Now, from the behavior of the function $K_\nu(z)$ for small values of its argument, $K_\nu(z) \sim \frac{1}{2}\Gamma(\nu)(z/2)^{-\nu}$, $z \to 0$, we get, in the case when M is small,

$$\rho_\phi = -\frac{1}{a^p b^{q+1}} \left\{ M K_1 \left(\frac{2\pi a}{b} M \right) + \sum_{h=0}^{p} \binom{p}{h} 2^h \sum_{n_h=1}^{\infty} \frac{M}{\sqrt{\sum_{j=1}^{h} n_j^2}} \right.$$

$$\left. \cdot K_1 \left(\frac{2\pi a}{b} M \sqrt{\sum_{j=1}^{h} n_j^2} \right) + \mathcal{O} \left[q\sqrt{1+M^2} K_1 \left(\frac{2\pi a}{b} \sqrt{1+M^2} \right) \right] \right\}. \tag{10.10}$$

The only presence of the mass-dimensional parameter μ is as M/μ everywhere, and this does not affect the small-M limit, $M/\mu \ll b/a$. Inserting back the $\hbar$ and c factors, we get

$$\rho_\phi = -\frac{\hbar c}{2\pi a^{p+1} b^q} \left[1 + \sum_{h=0}^{p} \binom{p}{h} 2^h \alpha \right] + \mathcal{O} \left[q K_1 \left(\frac{2\pi a}{b} \right) \right], \tag{10.11}$$

where α is a computable finite constant, obtained as an explicit geometrical sum in the limit $M \to 0$. It is remarkable that we do get a well defined limit, independent of M^2, provided M^2 is small enough.[15]

10.4.2 Numerical Results

For the most common cases, the constant α in (10.11) has been calculated to be of order 10^2, and the whole factor, in brackets, of order 10^7. This clearly shows the value of a precise calculation, as the one undertaken here, together with the fact that just a naive consideration of the dependencies of ρ_ϕ on the powers of the compactification radii, a and b, is actually *not enough* in order to get the correct result. Note, moreover, the non-trivial change in the power dependencies on going from (10.10) to (10.11).

---

[15] Indeed, a physically nice situation turns out to correspond to the mathematically rigorous case.

Table 8 The vacuum energy density contribution, in units of erg/cm^3 [see (10.3)]. In brackets, the values that more exactly match the one for the cosmological constant coming from observations, and in parenthesis the otherwise closest approximations

ρ_ϕ	$p=0$	$p=1$	$p=2$	$p=3$
$b=l_P$	10^{-13}	10^{-6}	1	10^5
$b=10l_P$	10^{-14}	$[10^{-8}]$	10^{-3}	10
$b=10^2 l_P$	10^{-15}	(10^{-10})	10^{-6}	10^{-3}
$b=10^3 l_P$	10^{-16}	10^{-12}	$[10^{-9}]$	(10^{-7})
$b=10^4 l_P$	10^{-17}	10^{-14}	10^{-12}	10^{-11}
$b=10^5 l_P$	10^{-18}	10^{-16}	10^{-15}	10^{-15}

Naturally enough, for the compactification radii at small scales, b, we take the Planck length, $b \sim l_{Planck}$, and for the large scales, a, the present size of the universe, $a \sim R_U$. With these choices, the order of a/b in the argument of K_1 is as big as: $a/b \sim 10^{60}$.[16] The final expression for the vacuum energy density is independent of the mass M of the field, provided this is small enough (eventually zero). In fact, the last term in (10.11) is exponentially vanishing (zero, for *app*). In ordinary units the bound on the mass of the scalar field is $M \leq 1.2 \times 10^{-32}$ eV (e.g., physically zero, since it is less by several orders of magnitude than any bound coming from SUSY theories).[17]

By replacing such values we obtain Table 8. The total number of large space dimensions is three (our universe). Good coincidence with the observational value is obtained for p large and $q = p + 1$ small compactified dimensions, $p = 0, \ldots, 3$, and this for the small compactification length, b, of the order of 10 to 10^3 times the Planck length l_P (a most reasonable range, according to string theory models). The p large and q small dimensions are *not* all that are supposed to exist: p and q refer to the *compactified* ones only. There may be non-compactified dimensions, what translates into a modification of the formulas above, but does not change the *order of magnitude* of the final numbers (see e.g. [10, 11, 14] for an elaboration on this technical point). Finally, simple power counting is *unable* to provide the correct order of magnitude of the results here obtained. One should however observe that the sign of the cc is a problem with this oversimplified models (which commonly get it wrong!). This is no longer so with the more elaborate theories to be considered below (see also [319, 357–365]).

10.5 Braneworld Models

Braneworld theories may help to solve both the hierarchy problem and the cc problem. And the bulk Casimir effect can play an important role in the construction

---

[16]Note that the square of this value yields the 120 orders of magnitude of the QFT cc.

[17]Where in fact scalar fields with low masses of the order of that of the lightest neutrino do show up [333], which may have observable implications.

(radion stabilization) of braneworlds. We have calculated the bulk Casimir effect (effective potential) for conformal and for massive scalar fields [366, 367]. The bulk is a 5-dim AdS or dS space, with 2 (or 1) 4-dim dS branes (our universe). The results obtained are consistent with observational data. We can only present a summary of those results here.

For the case of two dS_4 branes (at L separation) in a dS_5 background (it becomes a one-brane configuration as $L \to \infty$) the Casimir energy density and effective potential, for a conformally invariant scalar-gravitational theory $S = \frac{1}{2} \int d^5x \sqrt{g} [-g^{\mu\nu}\partial_\mu\phi\partial_\nu\phi + \xi_5 R^{(5)}\phi^2]$, $\xi_5 = -3/16$, with $R^{(5)}$ the curvature and $ds^2 = g_{\mu\nu} dx^\mu dx^\nu = \frac{\alpha^2}{\sinh^2 z}(dz^2 + d\Omega_4^2)$ the Euclidean metric of the 5-dim AdS bulk, $d\Omega_4^2 = d\xi^2 + \sin^2\xi\, d\Omega_3^2$—with α the AdS radius, related to the cc of the AdS bulk, and $d\Omega_3$ the metric on the 3-sphere being R the radius—are obtained as follows. For the one-brane Casimir energy density (pressure), we get [366]

$$\mathcal{E}_{Cas} = \frac{\hbar c}{2LR^4}\zeta\left(-\frac{1}{2}\Big|L_5\right) = -\frac{\hbar c\pi^3}{36L^6}\left[\frac{\pi^2}{315} - \frac{1}{240}\left(\frac{L}{R}\right)^2 + \mathcal{O}\left(\frac{L}{R}\right)^4\right], \quad (10.12)$$

which is about ten times larger than the ordinary Casimir effect: $\mathcal{E}_{CE} = -\frac{\hbar c\pi^2}{240L^4}$ (about 100 dynes/cm^2 at 100 nm). For the one-loop effective potential, we obtain

$$V = \frac{1}{2L\,\mathrm{Vol}(M_4)} \log\det(L_5/\mu^2), \quad (10.13)$$

where $L_5 = -\partial_z^2 - \Delta^{(4)} - \xi_5 R^{(4)} = L_1 + L_4$, and $\log\det L_5 = \sum_{n,\alpha}\log(\lambda_n^2 + \lambda_\alpha^2) = -\zeta'(0|L_5)$. In the one-brane limit $L \to \infty$, $K_t(L_1) \sim \frac{L}{2\sqrt{\pi t}}$ and $\zeta'(0|L_5) = \frac{1}{3R}[\zeta_H(-4, \frac{3}{2}) - \frac{1}{4}\zeta_H(-2, \frac{3}{2})] = 0$. And the small distance expansion for the effective potential yields (up to an overall factor)

$$\zeta'(0|L_5) = \frac{\zeta'(-4)}{6}\frac{\pi^4\mathcal{R}^4}{L^4} + \frac{\zeta'(-2)}{12}\frac{\pi^2\mathcal{R}^2}{L^2}$$

$$+ \frac{1}{24}\left[\zeta_H'(-4, 3/2) - \frac{1}{2}\zeta_H'(-2, 3/2)\right]\ln\frac{\pi^2\mathcal{R}^2}{L^2}$$

$$+ \frac{\zeta'(0)}{6}\left[\zeta_H'(-4, 3/2) - \frac{1}{2}\zeta_H'(-2, 3/2)\right] + \frac{1}{24}\zeta_H'(-4, 3/2)$$

$$+ \frac{1}{36}\left[\frac{1}{8}\zeta_H'(-4, 3/2) - \frac{1}{3}\zeta_H'(-6, 3/2)\right]\frac{L^2}{\mathcal{R}^2} + \mathcal{O}\left(\frac{L^4}{\pi^4\mathcal{R}^4}\right)$$

$$\simeq 0.129652\frac{\mathcal{R}^4}{L^4} - 0.025039\frac{\mathcal{R}^2}{L^2} - 0.002951\ln\frac{\mathcal{R}^2}{L^2} - 0.017956$$

$$- 0.000315\frac{L^2}{\mathcal{R}^2} + \cdots. \quad (10.14)$$

On the other hand, the effective potential for the massive scalar field model is obtained to be [366]

$$V = \frac{1}{2L\,\mathrm{Vol}(M_4)}\log\det\left(L_5/\mu^2\right),$$

$$L_5 \equiv -\partial_z^2 + m^2 l^2 \sinh^{-2} z - \Delta^{(4)} - \xi_5 R^{(4)} = L_1 + L_4 \quad (AdS), \quad (10.15)$$

$$L_5 \equiv -\partial_z^2 + m^2 \cosh^{-2} z - \Delta^{(4)} - \xi_5 R^{(4)} = L_1 + L_4 \quad (dS).$$

For the small mass limit (with L not large), it yields

$$\zeta'(0|L_5) \simeq \frac{a\rho + a^2\rho^2}{48} - \frac{\pi^2}{144}\left\{\frac{a\rho^2}{2} + \left[2\zeta'(-4,3/2) - \zeta'(-2,3/2)\right]\rho\right\}$$

$$- \frac{\pi^4}{4370}\left[2\zeta'(-4,3/2) - \zeta'(-2,3/2)\right]\rho^2 + \mathcal{O}(m^6), \qquad (10.16)$$

$$a \equiv \frac{\pi^2 \mathcal{R}^2}{L^2}, \qquad \rho \equiv \frac{m^2 l^2}{\pi^2}\frac{\tanh(L/2l)}{L/2l},$$

while for the large mass limit (with L not small), it is

$$\zeta'(0|L_5) = -\frac{4m^2 l^3}{3\mathcal{R}}\frac{\arctan(\sinh L/2l)}{\sinh(L/2l)} + \cdots, \qquad (10.17)$$

which is now non-zero (unlike in previous calculations, which turned a vanishing value) and can fit the observed order of magnitude under appropriate conditions.

10.6 Supergraviton Theories

We have also computed the effective potential for some multi-graviton models with supersymmetry [368–371]. In one case, the bulk is a flat manifold with the torus topology $\mathbb{R} \times \mathbb{T}^3$, and it can be shown that the induced cosmological constant can be rendered *positive* due to topological contributions [372, 373]. Previously, the case of $\mathbb{R}^4$ had been considered. In the multi-graviton model the induced cosmological constant can indeed be positive, but only if the number of massive gravitons is sufficiently large, what is not easy to fit in a natural way. In the supersymmetric case, however, the cosmological constant turns out to be positive just by imposing anti-periodic BC in the fermionic sector. An essential issue in our model is to allow for non-nearest-neighbor couplings.

The multi-graviton model is defined by taking N-copies of the fields with graviton $h_{n\mu\nu}$ and Stückelberg fields $A_{n\mu}$ and φ_n. Our theory is defined by a Lagrangian which is a generalization of the one in [374]. It reads

$$\mathcal{L} = \sum_{n=0}^{N-1}\left[-\frac{1}{2}\partial_\lambda h_{n\mu\nu}\partial^\lambda h_n^{\mu\nu} + \partial_\lambda h_{n\mu}^\lambda \partial_\nu h_n^{\mu\nu} - \partial_\mu h_n^{\mu\nu}\partial_\nu h_n + \frac{1}{2}\partial_\lambda h_n \partial^\lambda h_n\right.$$

$$-\frac{1}{2}\big(m^2 \Delta h_{n\mu\nu}\Delta h_n^{\mu\nu} - (\Delta h_n)^2\big) - 2\big(m\Delta^\dagger A_n^\mu + \partial^\mu \varphi_n\big)\big(\partial^\nu h_{n\mu\nu} - \partial_\mu h_n\big)$$

$$-\frac{1}{2}(\partial_\mu A_{n\nu} - \partial_\nu A_{n\mu})\big(\partial^\mu A_n^\nu - \partial^\nu A_n^\mu\big)\bigg]. \tag{10.18}$$

The Δ and $\Delta^\dagger$ are difference operators, which operate on the indices n as $\Delta\phi_n \equiv \sum_{k=0}^{N-1} a_k \phi_{n+k}$, $\Delta^\dagger \phi_n \equiv \sum_{k=0}^{N-1} a_k \phi_{n-k}$, $\sum_{k=0}^{N-1} a_k = 0$, where the a_k are N constants and the N variables ϕ_n can be identified with periodic fields on a lattice with N sites if the periodic boundary conditions, $\phi_{n+N} = \phi_n$, are imposed. The latter condition assures that Δ becomes the usual differentiation operator in a properly defined continuum limit.

In the case when anti-periodic boundary conditions are imposed in the fermionic sector, the situation changes completely with respect to the bosonic one, since the fermionic mass spectrum becomes quite different. The one-loop effective potential in the anti-periodic case is calculated to be

$$\begin{aligned}
V_{\mathit{eff}} &= \frac{M_1^4}{4\pi^2}\left(\ln\frac{M_1^2}{\mu_R^2} - \frac{3}{2}\right) - \frac{4M_1^4}{3\pi^2}\int_1^\infty du\, G(M_1 ru)\big(u^2 - 1\big)^{3/2}\\
&\quad - \frac{\tilde{M}_0^4}{4\pi^2}\left(\ln\frac{\tilde{M}_0^2}{\mu_R^2} - \frac{3}{2}\right) + \frac{4\tilde{M}_0^4}{3\pi^2}\int_1^\infty du\, G(\tilde{M}_0 ru)\big(u^2 - 1\big)^{3/2}\\
&\quad - \frac{\tilde{M}_1^4}{8\pi^2}\left(\ln\frac{\tilde{M}_1^2}{\mu_R^2} - \frac{3}{2}\right) + \frac{2\tilde{M}_1^4}{3\pi^2}\int_1^\infty du\, G(\tilde{M}_1 ru)\big(u^2 - 1\big)^{3/2}\\
&= -\frac{m^4}{36\pi^2}\log\frac{2^{16}}{3^9} + V_T, \tag{10.19}
\end{aligned}$$

where V_T is the sum of all the topological contributions. Note that the first term on the rhs is always negative, but the whole effective potential can be positive, due to the presence of the topological term. Thus, in the regime $mr \ll 1$ one has

$$V_T \sim \frac{1}{8\pi^2 r^4} \quad \Longrightarrow \quad V_{\mathit{eff}} > 0 \quad \text{for } mr < \left(\frac{2}{9}\log\frac{2^{16}}{3^9}\right)^{-1/4} \sim 1.4, \tag{10.20}$$

while in the opposite regime, $mr \gg 1$, we can see that the topological contribution (although still positive) is negligible, and the effective potential remains negative. In Fig. 14, the corresponding plot of the full effective potential, (10.19), is depicted as a function of $y \equiv mr$. The change of sign in the correct region is clearly observed.

To summarize, in the case of the torus topology we have obtained that the topological contributions to the effective potential have always a fixed sign, which depends on the BC one imposes. They are negative for periodic fields, and positive for anti-periodic fields. But topology provides then a mechanism which, in a most natural way, permits to have a positive cc in the multi-supergravity model with anti-periodic fermions. Moreover, the value of the cc is regulated by the corresponding size of the torus. We can most naturally use the minimum number, $N = 3$, of copies

Fig. 14 Plot of
$\tilde{V}_{eff}(y) \equiv r^4 V_{eff}(r)$, (10.19),
as a function of $y \equiv mr$

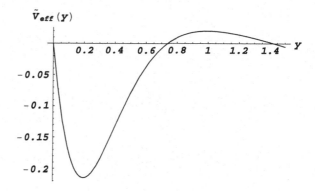

of bosons and fermions, and show that—as in the first, much more simple example, but now with the right sign!—within our model the observational values for the cosmological constant, (10.3), can be matched, by making very reasonable adjustments of the parameters involved. As a byproduct, the results that we have obtained [372, 373] may also be relevant in the study of electroweak symmetry breaking in models with similar type of couplings, for the deconstruction issue.

References

1. C. Reid, *Hilbert* (Springer, Berlin, 1970), p. 82
2. P. Epstein, Math. Ann. **56**, 615 (1903)
3. A. Erdélyi, W. Magnus, F. Oberhettinger, F.G. Tricomi, *'Bateman Manuscript Project', Higher Transcendental Functions* (McGraw-Hill, New York, 1953)
4. P. Epstein, Math. Ann. **65**, 205 (1907)
5. S. Iyanaga, Y. Kawada (eds.), *Encyclopedic Dictionary of Mathematics*, vol. II (MIT Press, Cambridge, 1977), p. 1372 ff
6. E.C. Titchmarsh, *The Zeta Function of Riemann* (Cambridge University Press, Cambridge, England, 1930)
7. H.M. Edwards, *Riemann's Zeta Function* (Academic Press, San Diego, 1974)
8. A. Ivić, *The Riemann Zeta Function* (Wiley, New York, 1985)
9. J. Jorgenson, S. Lang, *Basic Analysis of Regularized Series and Products*, Lecture Notes in Mathematics, vol. 1564 (Springer, Berlin, 1993)
10. E. Elizalde, S.D. Odintsov, A. Romeo, A.A. Bytsenko, S. Zerbini, *Zeta Regularization Techniques with Applications* (World Scientific, Singapore, 1994)
11. E. Elizalde, *Ten Physical Applications of Spectral Zeta Functions* (Springer, Berlin, 1995)
12. A.A. Karatsuba, S.M. Voronin, *The Riemann Zeta-Function* (de Gruyter, Hawthorne, 1992)
13. K. Kirsten, *Spectral Functions in Mathematics and Physics* (Chapman & Hall, London, 2001)
14. A.A. Bytsenko, G. Cognola, E. Elizalde, V. Moretti, S. Zerbini, *Analytic Aspects of Quantum Fields* (World Scientific, Singapore, 2003)
15. T.M. Apostol, Zeta and related functions, in *NIST Handbook of Mathematical Functions*, ed. by F.W.J. Olver et al. (Cambridge University Press, Cambridge, 2010)
16. G.H. Hardy, J.E. Littlewood, Contributions to the theory of the Riemann zeta-function and the theory of the distribution of primes. Acta Math. **41**, 119 (1916)
17. G.H. Hardy, Mess. Math. **49**, 85 (1919)
18. G.H. Hardy, *Divergent Series* (Clarendon Press, Oxford, 1949)
19. T. Carleman, Propriétés asymptotiques des fonctions fondamentales des membranes vibrantes (French), Skand. Mat.-Kongr. **8**, 34–44 (1935)
20. S. Minakshisundaram, Å. Pleijel, Can. J. Math. **1**, 242 (1949)
21. R.T. Seeley, Complex powers of an elliptic operator: Singular integrals, in *Proc. Symp. Pure Math.*, Chicago, IL, 1966 (Am. Math. Soc., Providence, 1967), pp. 288–307
22. D.B. Ray, I.M. Singer, Adv. Math. **7**, 145 (1971)
23. J.S. Dowker, R. Critchley, Phys. Rev. D **13**, 3224 (1976)
24. S.W. Hawking, Commun. Math. Phys. **55**, 133 (1977)
25. S. Chowla, A. Selberg, Proc. Natl. Acad. Sci. USA **35**, 317 (1949)
26. R.T. Seeley, Am. J. Math. **91**, 889 (1969)

E. Elizalde, *Ten Physical Applications of Spectral Zeta Functions*,
Lecture Notes in Physics 855,
DOI 10.1007/978-3-642-29405-1, © Springer-Verlag Berlin Heidelberg 2012

27. B.S. De Witt, Phys. Rep. **19**, 295 (1975)
28. H.P. McKean, I.M. Singer, J. Differ. Geom. **1**, 43 (1967)
29. P. Gilkey, J. Differ. Geom. **10**, 601 (1975)
30. P. Gilkey, Am. Math. Soc. Proc. Symp. Pure Math. **27**, 265 (1975)
31. P. Gilkey, Compos. Math. **38**, 201 (1979)
32. P. Gilkey, Contemp. Math. **73**, 79 (1988)
33. E. Elizalde, S. Leseduarte, S. Zerbini, Mellin transform techniques for zeta-function resummations, UB-ECM-PF 93/7, arXiv: hep-th/9303126 (1993)
34. P. Candelas, P. Raine, Phys. Rev. D **12**, 965 (1975)
35. A. Actor, Class. Quantum Gravity **5**, 1415 (1988)
36. I. Brevik, H.B. Nielsen, Phys. Rev. D **41**, 1185 (1990)
37. X. Li, X. Shi, J. Zhang, Phys. Rev. D **44**, 560 (1991)
38. B.F. Svaiter, N.F. Svaiter, Phys. Rev. D **47**, 4581 (1993)
39. L. Brink, H.B. Nielsen, Phys. Lett. B **45**, 332 (1973)
40. A. Salam, J. Strathdee, Nucl. Phys. B **90**, 203 (1975)
41. L.S. Brown, G.J. MacLay, Phys. Rev. **184**, 1272 (1969)
42. S.K. Blau, M. Visser, A. Wipf, Nucl. Phys. B **310**, 163 (1988)
43. E. Elizalde, K. Kirsten, J. Math. Phys. **35**, 1260 (1994)
44. D.G.C. McKeon, T.N. Sherry, Phys. Rev. Lett. **59**, 532 (1987)
45. A. Rebhan, Phys. Rev. D **39**, 3101 (1989)
46. A.Y. Shiekh, Can. J. Phys. **68**, 620 (1990)
47. D.G.C. McKeon, T.N. Sherry, Phys. Rev. D **35**, 3854 (1987)
48. E. Elizalde, A. Romeo, Phys. Rev. D **40**, 436 (1989)
49. V.M. Mostepanenko, N.N. Trunov, *Casimir Effect and Its Applications* (in Russian) (Energoatomizdat, Moscow, 1990)
50. A. Antillón, G. Germán, Phys. Rev. D **47**, 4567 (1993)
51. E. Elizalde, J. Math. Phys. **31**, 170 (1990)
52. I. Brevik, I. Clausen, Phys. Rev. D **39**, 603 (1989)
53. A. Actor, J. Phys. A **24**, 3741 (1991)
54. D.M. McAvity, H. Osborn, Nucl. Phys. B **394**, 728 (1993)
55. S.A. Frolov, A.A. Slavnov, Phys. Lett. B **309**, 344 (1993)
56. M. Bordag, E. Elizalde, K. Kirsten, J. Math. Phys. **37**, 895 (1996)
57. I.G. Avramidi, Nucl. Phys. B **355**, 712 (1991)
58. S.A. Fulling, G. Kennedy, Trans. Am. Math. Soc. **310**, 583 (1988)
59. P. Amsterdamski, A.L. Berkin, D.J. O'Connor, Class. Quantum Gravity **6**, 1981 (1989)
60. A. Voros, Commun. Math. Phys. **110**, 439 (1987)
61. P. Ramond, *Field Theory: A Modern Primer* (Benjamin-Cummings, Reading, 1981)
62. N. Birrell, P.C.W. Davies, *Quantum Fields in Curved Spaces* (Cambridge University Press, Cambridge, 1982)
63. R. Forman, Commun. Math. Phys. **147**, 485 (1992)
64. M. Bordag, E. Elizalde, B. Geyer, K. Kirsten, Commun. Math. Phys. **179**, 215 (1996)
65. J.S. Dowker, Commun. Math. Phys. **162**, 633 (1994)
66. J.S. Dowker, Class. Quantum Gravity **11**, 557 (1994)
67. J.S. Dowker, J. Math. Phys. **35**, 4989 (1994); Erratum, J. Math. Phys. **36**, 988 (1995)
68. M. Kontsevich, S. Vishik, in *Functional Analysis on the Eve of the 21st Century*, vol. 1 (1993), pp. 173–197
69. D.B. Ray, Adv. Math. **4**, 109 (1970)
70. D.B. Ray, I.M. Singer, Ann. Math. **98**, 154 (1973)
71. R. Killip, Spectral theory via sum rules, in *Spectral Theory and Mathematical Physics*, Proc. Symp. Pure Math., vol. 76 (2007), p. 907
72. A. Connes, *Noncommutative Geometry* (Academic Press, New York, 1994)
73. M. Wodzicki, Noncommutative residue, Chapter I, in *Lecture Notes in Mathematics*, vol. 1289, ed. by Yu.I. Manin (Springer, Berlin, 1987), p. 320
74. E. Elizalde, J. Phys. A **30**, 2735 (1997)

75. C. Kassel, Asterisque **177**, 199 (1989), Sem. Bourbaki
76. P. Ramond, *Field Theory: A Modern Primer* (Addison-Wesley, Redwood City, 1989)
77. N. Evans, Phys. Lett. B **457**, 127 (1999)
78. J.S. Dowker, On the relevance of the multiplicative anomaly, arXiv: hep-th/9803200 (1998)
79. E. Elizalde, A. Filippi, L. Vanzo, S. Zerbini, Is the multiplicative anomaly dependent on the regularization? arXiv: hep-th/9804071 (1998)
80. E. Elizalde, A. Filippi, L. Vanzo, S. Zerbini, Is the multiplicative anomaly relevant? arXiv: hep-th/9804072 (1998)
81. E. Elizalde, G. Cognola, S. Zerbini, Nucl. Phys. B **532**, 407 (1998)
82. J.J. McKenzie-Smith, D.J. Toms, Phys. Rev. D **58**, 105001 (1998)
83. A.A. Bytsenko, F.L. Williams, J. Math. Phys. **39**, 1075 (1998)
84. C.M. Bender, S.A. Orszag, *Advanced Mathematical Methods for Scientists and Engineers* (McGraw-Hill, New York, 1978), pp. 261–265
85. S. Chowla, A. Selberg, Proc. Natl. Acad. Sci. USA **35**, 317 (1949)
86. E. Elizalde, J. Phys. A **27**, 3775 (1994)
87. E. Elizalde, Commun. Math. Phys. **198**, 83 (1998)
88. D.G.C. McKeon, T.N. Sherry, Phys. Rev. Lett. **59**, 532 (1987)
89. D.G.C. McKeon, T.N. Sherry, Phys. Rev. D **35**, 3854 (1987)
90. E. Elizalde, S. Naftulin, S.D. Odintsov, Phys. Rev. D **49**, 2852 (1994)
91. R.B. Mann, L. Tarasov, D.G.C. McKeon, T. Steele, Nucl. Phys. B **311**, 630 (1989)
92. A.Y. Shiekh, Can. J. Phys. **74**, 172 (1996)
93. A. Rebhan, Phys. Rev. D **39**, 3101 (1989)
94. L. Culumovic, M. Leblanc, R.B. Mann, D.G.C. McKeon, T.N. Sherry, Phys. Rev. D **41**, 514 (1990)
95. E. Elizalde, Math. Comput. **47**, 347 (1986)
96. F. Steiner, Phys. Lett. B **188**, 447 (1987)
97. S. Rudaz, J. Math. Phys. **31**, 2832 (1990)
98. E. Elizalde, J. Math. Phys. **34**, 3222 (1993)
99. F.W.J. Olver, *Asymptotics and Special Functions* (Academic Press, New York, 1974), pp. 71–72 and 80–84
100. A. Erdélyi (ed.), *Higher Transcendental Functions*, vol. I (McGraw-Hill, New York, 1953), p. 27
101. A. Actor, Nucl. Phys. B **265**, 689 (1986)
102. H.A. Weldon, Nucl. Phys. B **270**, 79 (1986)
103. A. Actor, J. Phys. A **20**, 927 (1987)
104. A. Actor, J. Phys. A **20**, 5351 (1987)
105. I.J. Zucker, J. Phys. A **7**, 1568 (1974)
106. I.J. Zucker, J. Phys. A **8**, 1734 (1975)
107. I.J. Zucker, M.M. Robertson, J. Phys. A **8**, 874 (1975)
108. A. Actor, Fortschr. Phys. **35**, 793 (1987)
109. E. Elizalde, J. Phys. A **22**, 931 (1989)
110. E. Elizalde, A. Romeo, Int. J. Mod. Phys. A **5**, 1653 (1990)
111. J.J. Duistermaat, V.W. Guillemin, Invent. Math. **29**, 39 (1975)
112. N. Bleistein, R.A. Handelsman, *Asymptotic Expansions of Integrals* (Dover, New York, 1975)
113. S.A. Molchanov, Russian Math. Surveys **30**, 1 (1975)
114. J. Polchinski, Phys. Rev. Lett. **68**, 1267 (1992)
115. J. Polchinski, Phys. Rev. D **46**, 3667 (1992)
116. E. Elizalde, S. Leseduarte, S.D. Odintsov, Phys. Rev. D **48**, 1757 (1993)
117. H. Heilbronn, Q. J. Math. Oxford **5**, 150 (1934)
118. M. Deuring, Math. Z. **37**, 403 (1933)
119. E.T. Wittaker, G.N. Watson, *A Course of Modern Analysis*, 4th edn. (Cambridge University Press, Cambridge, 1965)

120. E. Elizalde, Zeta-function regularization techniques for series summations and applications, in *Proceedings of the Leipzig Workshop: Quantum Field Theory under the Influence of External Conditions*, 1992

121. T. Landsberg, J. Math. **CXI**, 234 (1893)

122. E. Elizalde, J. Math. Phys. **35**, 6100 (1994)

123. E. Elizalde, Commun. Math. Phys. **198**, 83 (1998)

124. E. Elizalde, J. Phys. A **30**, 2735 (1997)

125. E. Elizalde, J. Comput. Appl. Math. **118**, 125 (2000)

126. E. Elizalde, J. Math. Phys. **35**, 3308 (1994)

127. E. Elizalde, Yu. Kubyshin, J. Phys. A **27**, 7533 (1994)

128. H.B.G. Casimir, Proc. K. Ned. Akad. Wet. B **51**, 793 (1948)

129. H.B.G. Casimir, Physica **19**, 846 (1953)

130. G. Plunien, B. Müller, W. Greiner, Phys. Rep. **134**, 87 (1986)

131. Yu.S. Barash, V.L. Ginzburg, Electromagnetic fluctuations and molecular forces in condensed matter, in *The Dielectric Function of Condensed Systems*, ed. by L.V. Keldysh et al. (Elsevier, Amsterdam, 1989), pp. 389–457

132. H.B.G. Casimir, D. Polder, Phys. Rev. **73**, 360 (1948)

133. E. Elizalde, A. Romeo, Am. J. Phys. **59**, 711 (1991)

134. D. van der Waals, *Die Kontinuität des Garformigen und Flüssigen Zuständes* (Amsterdam, 1881)

135. F. London, Z. Phys. **63**, 245 (1930)

136. E.M. Lifshitz, Ž. Èksp. Teor. Fiz. **29**, 94 (1955) [Sov. Phys. JETP **2**, 73 (1956)]

137. J. Bernabéu, R. Tarrach, Ann. Phys. (N.Y.) **102**, 323 (1976)

138. I.E. Dzyaloshinskii, E.M. Lifshitz, L.P. Pitaevskii, Adv. Phys. **10**, 165 (1961)

139. V.A. Parsegian, B.W. Ninham, J. Chem. Phys. **52**, 4578 (1970)

140. V.A. Parsegian, B.W. Ninham, Biophys. J. **10**, 646 (1970)

141. V.A. Parsegian, B.W. Ninham, Biophys. J. **10**, 664 (1970)

142. F. Sauer, Dissertation (Universität Göttingen, 1992)

143. J. Mehra, Physica **37**, 145 (1967)

144. S.K. Lamoreaux, Phys. Rev. Lett. **78**, 5 (1997); Erratum, Phys. Rev. Lett. **81**, 5475 (1998)

145. https://www.cfa.harvard.edu/~babb/casimir-bib.html

146. E. Elizalde, S.D. Odintsov, A.A. Saharian, Phys. Rev. D **79**, 065023 (2009)

147. E. Elizalde, A.A. Saharian, T.A. Vardanyan, Phys. Rev. D **81**, 124003 (2010)

148. J. Ambjørn, S. Wolfram, Ann. Phys. (N.Y.) **147**, 1 (1983)

149. D. Dalvit et al., *Casimir Physics*, Lecture Notes in Physics, vol. 834 (Springer, Berlin, 2011)

150. M. Bordag et al., *Advances in the Casimir Effect*, International Series of Monographs on Physics, vol. 145 (Oxford Science, Oxford, 2009)

151. K.A. Milton et al., *The Casimir Effect: Physical Manifestations of Zero-Point Energy* (World Scientific, UK, 2001)

152. V. Mostepanenko, N.N. Trunov, *The Casimir Effect and Its Applications* (Oxford Science, Oxford, 1997)

153. P.W. Milonni, *The Quantum Vacuum: An Introduction to Quantum Electrodynamics* (Academic Press, New York, 1994)

154. M. Krech, *The Casimir Effect in Critical Systems* (World Scientific, UK, 1994)

155. I. Brevik, H. Skurdal, R. Sollie, J. Phys. A **27**, 6853 (1994)

156. E. Cheng, M.W. Cole, W.F. Saam, J. Treiner, Phys. Rev. Lett. **67**, 1007 (1991)

157. M. Krech, S. Dietrich, Phys. Rev. Lett. **66**, 345 (1991)

158. M. Krech, S. Dietrich, Phys. Rev. Lett. **67**, 1055 (1991)

159. M. Krech, S. Dietrich, Phys. Rev. A **46**, 1886 (1992)

160. M. Krech, S. Dietrich, Phys. Rev. A **46**, 1922 (1992)

161. F. De Martini, G. Jacobovitz, Phys. Rev. Lett. **60**, 1711 (1988)

162. F. De Martini et al., Phys. Rev. A **43**, 2480 (1991)

163. W.I. Weisberger, Commun. Math. Phys. **112**, 633 (1987)

164. W.I. Weisberger, Nucl. Phys. B **284**, 171 (1987)

165. E. Aurell, P. Salomonson, Commun. Math. Phys. **165**, 233 (1994)
166. C. Itzykson, J.-M. Luck, J. Phys. A **19**, 211 (1986)
167. J.S. Dowker, R. Banach, J. Phys. A **11**, 2255 (1978)
168. P. Candelas, S. Weinberg, Nucl. Phys. B **237**, 397 (1984)
169. G.W. Gibbons, J. Phys. A **11**, 1341 (1978)
170. E.J. Copeland, D.J. Toms, Nucl. Phys. B **255**, 201 (1985)
171. G.R. Shore, Ann. Phys. **128**, 376 (1980)
172. N.N. Bogoliubov, D.V. Shirkov, *Introduction to the Theory of Quantized Fields* (Wiley, New York, 1980)
173. I.L. Buchbinder, S.D. Odintsov, I.L. Shapiro, *Effective Action in Quantum Gravity* (IOP Publishing, Boston and Philadelphia, 1992)
174. T. Inagaki, T. Kouno, T. Muta, Int. J. Mod. Phys. A **10**, 2241 (1995)
175. S. Carlip, Class. Quantum Gravity **11**, 31 (1994)
176. S. Lang, *Elliptic Functions*, Grad. Text in Mathematics, vol. 112 (Springer, New York, 1987), Chap. 20
177. T. Kubota, *Elementary Theory of Eisenstein Series* (Kodansha, Tokyo, and Wiley, New York, 1973)
178. A. Connes, M.R. Douglas, A. Schwarz, JHEP **9802**, 003 (1998)
179. M.R. Douglas, C. Hall, JHEP **9802**, 008 (1998)
180. N. Seiberg, E. Witten, JHEP **9909**, 032 (1999)
181. Y.-K.E. Cheung, M. Krogh, Nucl. Phys. B **528**, 185 (1998)
182. C.-S. Chu, P.-M. Ho, Nucl. Phys. B **550**, 151 (1999)
183. V. Schomerus, JHEP **9906**, 030 (1999)
184. F. Ardalan, H. Arfaei, M.M. Sheikh-Jabbari, JHEP **9902**, 016 (1999)
185. A.A. Bytsenko, A.E. Goncalves, S. Zerbini, Mod. Phys. Lett. A **16**, 1479 (2001)
186. P.B. Gilkey, *Invariance Theory, the Heat Equation, and the Atiyah–Singer Index Theorem*, Studies in Advanced Mathematics (CRC Press, Boca Raton, 1995)
187. A.A. Bytsenko, E. Elizalde, S. Zerbini, Phys. Rev. D **64**, 105024 (2001)
188. E. Elizalde, Phys. Lett. B **342**, 277 (1995)
189. R. Narayanan, H. Neuberger, Phys. Lett. B **302**, 62 (1993)
190. S. Aoki, Y. Kikukawa, Mod. Phys. Lett. A **8**, 3517 (1993)
191. K. Fujikawa, Nucl. Phys. B **428**, 169 (1994)
192. K. Fujikawa, Phys. Rev. D **21**, 2848 (1980); Erratum, Phys. Rev. D **22**, 1499 (1980)
193. K. Fujikawa, Phys. Rev. Lett. **42**, 1195 (1979)
194. E. Elizalde, J. Phys. A **27**, 3775 (1994)
195. I. Brevik, E. Elizalde, Phys. Rev. D **49**, 5319 (1994)
196. I. Brevik, H.B. Nielsen, Phys. Rev. D **51**, 1869 (1995)
197. T.D. Lee, *Particle Physics and Introduction to Field Theory* (Harwood Academic, New York, 1988), Chaps. 17 and 20. Media obeying the condition $\epsilon\mu = 1$ have been discussed also, for example, by I. Brevik, I. Clausen, Phys. Rev. D **39**, 603 (1989)
198. N.G. van Kampen, B.R.A. Nijboer, K. Schram, Phys. Lett. A **26**, 307 (1968)
199. P.M. Morse, H. Feshbach, *Methods of Theoretical Physics* (McGraw-Hill, New York, 1953), p. 428
200. E.M. Lifshitz, L.P. Pitaevskii, *Statistical Physics, Part 2* (Pergamon, Oxford, 1980)
201. E. Elizalde, J. Phys A **36**, L567 (2003)
202. E. Wigner, Commun. Pure Appl. Math. **13**, 1 (1960)
203. H.B.G. Casimir, Proc. K. Ned. Acad. Wet. **51**, 635 (1948)
204. E. Elizalde, J. Phys. A **34**, 3025 (2001)
205. E. Elizalde, J. Comput. Appl. Math. **118**, 125 (2000)
206. E. Elizalde, Commun. Math. Phys. **198**, 83 (1998)
207. E. Elizalde, J. Phys. A **30**, 2735 (1997)
208. K. Kirsten, E. Elizalde, Phys. Lett. B **365**, 72 (1995)
209. E. Elizalde, J. Phys. A **27**, 3775 (1994)
210. E. Elizalde, J. Phys. A **27**, L299 (1994)

211. E. Elizalde, J. Phys. A **22**, 931 (1989)

212. E. Elizalde, A. Romeo, Phys. Rev. D **40**, 436 (1989)

213. D. Deutsch, P. Candelas, Phys. Rev. D **20**, 3063 (1979)

214. K. Symanzik, Nucl. Phys. B **190**, 1 (1981)

215. R.L. Jaffe, Unnatural acts: Unphysical consequences of imposing boundary conditions on quantum fields, CTP-MIT-3394, arXiv: hep-th/0307014v2

216. N. Graham, R.L. Jaffe, V. Khemani, M. Quandt, M. Scandurra, H. Weigel, Phys. Lett. B **572**, 196 (2003)

217. N. Graham, R.L. Jaffe, V. Khemani, M. Quandt, M. Scandurra, H. Weigel, Nucl. Phys. B **645**, 49 (2002)

218. N. Graham, R.L. Jaffe, H. Weigel, Int. J. Mod. Phys. A **17**, 846 (2002)

219. V.M. Mostepanenko, N.N. Trunov, *The Casimir Effect and Its Application* (Clarendon Press, Oxford, 1997)

220. K.A. Milton, *The Casimir Effect: Physical Manifestations of Zero-Point Energy* (World Scientific, Singapore, 2001)

221. M. Bordag, U. Mohideen, V.M. Mostepanenko, Phys. Rep. **353**, 1 (2001)

222. E. Elizalde, M. Bordag, K. Kirsten, J. Phys. A **31**, 1743 (1998)

223. E. Elizalde, L. Vanzo, S. Zerbini, Commun. Math. Phys. **194**, 613 (1998)

224. M. Bordag, E. Elizalde, K. Kirsten, S. Leseduarte, Phys. Rev. D **56**, 4896 (1997)

225. M. Bordag, E. Elizalde, K. Kirsten, J. Math. Phys. **37**, 895 (1996)

226. M. Bordag, E. Elizalde, B. Geyer, K. Kirsten, Commun. Math. Phys. **179**, 215 (1996)

227. E. Elizalde, A.C. Tort, Phys. Rev. D **66**, 045033 (2002)

228. L. Blanchet, G. Faye, J. Math. Phys. **41**, 7675 (2000)

229. V. Moretti, Commun. Math. Phys. **232**, 189 (2003)

230. H. Sahlmann, R. Verch, Rev. Math. Phys. **13**, 1203 (2001)

231. R. Brunetti, K. Fredenhagen, Commun. Math. Phys. **208**, 623 (2000)

232. V. Moretti, J. Math. Phys. **40**, 3843 (1999)

233. R. Brunetti, K. Fredenhagen, Commun. Math. Phys. **180**, 633 (1996)

234. R. Verch, Hadamard vacua in curved spacetime and the principle of local definiteness, in *XIth International Congress of Mathematical Physics*, Paris, 1994 (International Press, Cambridge, 1995), p. 352

235. P.L. Butzer, A. Kilbas, J.J. Trujillo, J. Math. Anal. Appl. **269**, 387 (2002)

236. A.N. Guz, V.V. Zozulya, Int. J. Nonlinear Sci. Numer. Simul. **2**, 173 (2001)

237. R. Estrada, R. Kanwal, *Singular Integral Equations* (Birkhäuser Boston, Boston, 2000)

238. G. Criscuolo, J. Comput. Appl. Math. **78**, 255 (1997)

239. N. Mastronardi, D. Occorsio, J. Comput. Appl. Math. **70**, 75 (1996)

240. J.D. Elliott, J. Comput. Appl. Math. **62**, 267 (1995)

241. L. Schwartz, *Théorie des distributions* (Hermann, Paris, 1997)

242. Eur. Phys. J. H **35**(3) (Springer, 2010)

243. R. Melrose, Problems of Class 18.155, MIT, Third Assignment, Fall Term (2001)

244. *Mathematica, Version 5* (Wolfram Research, Inc., Champaign, 2003)

245. E. Elizalde, S.D. Odintsov, Mod. Phys. Lett. A **26**, 2369 (1992)

246. E. Elizalde, S.D. Odintsov, Phys. Rev. D **47**, 2497 (1993)

247. G.A. Vilkovisky, Nucl. Phys. B **234**, 125 (1984)

248. B.S. De Witt, in *Architecture of Fundamental Interactions at Short Distances*, ed. by P. Ramond, R. Stora (Elsevier, Amsterdam, 1987)

249. A.M. Polyakov, Nucl. Phys. B **268**, 406 (1986)

250. T.L. Curtright, G.I. Ghandhour, C.K. Zachos, Phys. Rev. Lett. **57**, 799 (1986)

251. T.L. Curtright, G.I. Ghandhour, C.K. Zachos, Phys. Rev. D **34**, 3811 (1986)

252. F. Alonso, D. Espriu, Nucl. Phys. B **283**, 393 (1987)

253. E. Braaten, R.D. Pisarski, S.-M. Tze, Phys. Rev. Lett. **58**, 93 (1987)

254. H. Kleinert, Phys. Rev. Lett. **58**, 1915 (1987)

255. P. Olesen, S.-K. Yang, Nucl. Phys. B **283**, 74 (1987)

256. R. Pisarski, Phys. Rev. Lett. **58**, 1300 (1987)

257. E. Bergshoeff, E. Sezgin, P.K. Townsend, Ann. Phys. (N.Y.) **185**, 330 (1988)
258. M.J. Duff, Class. Quantum Gravity **6**, 1577 (1989)
259. A. Achúcarro, J.M. Evans, P.K. Townsend, D.L. Wiltshire, Phys. Lett. B **198**, 441 (1987)
260. M.J. Duff, T. Inami, C.N. Pope, E. Sezgin, K.S. Stelle, Nucl. Phys. B **297**, 515 (1988)
261. K. Fujikawa, J. Kubo, Phys. Lett. B **199**, 75 (1987)
262. K. Fujikawa, Phys. Lett. B **206**, 18 (1988)
263. E.G. Floratos, Phys. Lett. B **220**, 61 (1989)
264. E.G. Floratos, G.K. Leontaris, Phys. Lett. B **223**, 37 (1989)
265. S.D. Odintsov, D.L. Wiltshire, Class. Quantum Gravity **7**, 1499 (1990)
266. A.A. Bytsenko, S.D. Odintsov, Class. Quantum Gravity **9**, 391 (1992)
267. D.H. Hartley, M. Önder, R.W. Tucker, Class. Quantum Gravity **6**, 30 (1989)
268. A. Achúcarro, P. Kapusta, K.S. Stelle, Phys. Lett. B **232**, 302 (1989)
269. S.D. Odintsov, Phys. Lett. B **247**, 21 (1990)
270. E. Elizalde, A. Romeo, Int. J. Mod. Phys. A **7**, 7365 (1992)
271. Y. Nambu, in *Symmetries and Quark Models*, ed. by R. Chand (Gordon & Breach, New York, 1970)
272. T. Goto, Prog. Theor. Phys. **46**, 1560 (1971)
273. T. Eguchi, Phys. Rev. Lett. **44**, 126 (1980)
274. O. Alvarez, Phys. Rev. D **24**, 440 (1981)
275. E. Elizalde, Nuovo Cim. B **104**, 685 (1989)
276. C.W. Bernhard, Phys. Rev. D **9**, 3312 (1974)
277. L. Dolan, R. Jackiw, Phys. Rev. D **9**, 3320 (1974)
278. N.P. Landsman, Ch.G. van Weert, Phys. Rep. **145**, 141 (1987)
279. D.J. Gross, R.D. Pisarski, S. Rudaz, Rev. Mod. Phys. **53**, 43 (1981)
280. J.I. Kapusta, *Finite Temperature Field Theory* (Cambridge University Press, Cambridge, 1989)
281. G. Cognola, L. Vanzo, S. Zerbini, J. Math. Phys. **33**, 222 (1992)
282. F. Caruso, N.P. Neto, B.F. Svaiter, N.F. Svaiter, Phys. Rev. D **43**, 1300 (1991)
283. B.F. Svaiter, N.F. Svaiter, J. Math. Phys. **32**, 175 (1991)
284. B.P. Dolan, C. Nash, Commun. Math. Phys. **148**, 139 (1992)
285. A. Actor, Class. Quantum Gravity **7**, 663 (1990)
286. A. Actor, Class. Quantum Gravity **7**, 1463 (1990)
287. I.L. Buchbinder, S.D. Odintsov, Fortschr. Phys. **37**, 225 (1989)
288. L.H. Ford, T. Yoshimura, Phys. Lett. A **70**, 89 (1979)
289. L.H. Ford, Phys. Rev. D **21**, 933 (1980)
290. D.J. Toms, Phys. Rev. D **21**, 2805 (1980)
291. D.J. Toms, Phys. Rev. D **21**, 928 (1980)
292. G. Kennedy, Phys. Rev. D **23**, 2884 (1981)
293. S. Weinberg, Phys. Rev. D **9**, 3357 (1974)
294. A.D. Linde, Rep. Prog. Phys. **42**, 389 (1979)
295. Y.P. Goncharov, Phys. Lett. A **91**, 153 (1982)
296. G. Denardo, E. Spallucci, Nuovo Cim. A **64**, 27 (1981)
297. G. Denardo, E. Spallucci, Nucl. Phys. B **169**, 514 (1980)
298. E. Elizalde, A. Romeo, Phys. Lett. B **244**, 387 (1990)
299. A.D. Linde, Phys. Lett. B **123**, 185 (1983)
300. Y.P. Goncharov, Phys. Lett. B **147**, 269 (1984)
301. G.F.R. Ellis, Gen. Relativ. Gravit. **2**, 7 (1971)
302. G.F.R. Ellis, G. Schreiber, Phys. Lett. A **115**, 97 (1986)
303. H.V. Fagundes, Phys. Rev. Lett. **70**, 1579 (1993)
304. K. Kirsten, J. Math. Phys. **32**, 3008 (1991)
305. D.J. Toms, Phys. Lett. A **77**, 303 (1980)
306. R. Jaffe, Phys. Rev. D **72**, 021301 (2005)
307. S.A. Fulling et al., Phys. Rev. D **76**, 025004 (2007)
308. M. Tegmark et al. [SDSS Collaboration], Phys. Rev. D **69**, 103501 (2004)

309. D.J. Eisenstein et al., Astrophys. J. **633**, 560 (2005)
310. J.K. Adelman-McCarthy et al., Astrophys. J. Suppl. **162**, 38 (2006)
311. S. Weinberg, Rev. Mod. Phys. **61**, 1 (1989)
312. S. Weinberg, The cosmological constant problems (Talk given at Dark Matter 2000, February, 2000)
313. S.M. Carroll, Living Rev. Relativ. **4**, 1 (2001)
314. E. Baum, Phys. Lett. B **133**, 185 (1983)
315. S. Hawking, Phys. Lett. B **134**, 403 (1984)
316. S. Coleman, Nucl. Phys. B **310**, 643 (1988)
317. J. Polchinski, Phys. Lett. B **219**, 251 (1989)
318. E. Elizalde, Cosmology: Techniques and observations, in *Proceedings of the Second International Conference on Fundamental Interactions 2004*, Pedra Azul, ES, Brazil, ed. by M.C.B. Abdalla, A.A. Bytsenko, M.E.X. Guimarães, O. Piguet (Editora RiMa, São Paulo, Brazil, 2004), pp. 57–120
319. E. Elizalde, J. Phys. A **39**, 6299 (2006)
320. J. Haro, E. Elizalde, Phys. Rev. Lett. **97**, 130401 (2006)
321. M. Tegmark, Science **296**, 1427 (2002)
322. P. de Bernardis et al., Nature **404**, 955 (2000)
323. S. Hanany et al., Astrophys. J. **545**, L5 (2000)
324. A. Balbi et al., Astrophys. J. **545**, L1 (2000); Erratum, Astrophys. J. **558**, L145 (2001)
325. N.J. Cornish et al., Phys. Rev. Lett. **92**, 201302 (2004)
326. J.-P. Luminet et al., Nature **425**, 593 (2003)
327. S. Perlmutter et al. [Supernova Cosmology Project Collaboration], Astrophys. J. **517**, 565 (1999)
328. A.G. Riess et al. [Hi-Z Supernova Team Collaboration], Astron. J. **116**, 1009 (1998)
329. A.G. Riess, Publ. Astron. Soc. Pac. **112**, 1284 (2000)
330. S.M. Carroll, Living Rev. Relativ. **4**, 1 (2001)
331. S.M. Carroll, Why is the Universe accelerating? in *Contribution to Measuring and Modeling the Universe*, ed. by W.L. Freedman. Carnegie Observatories Astrophysics Series, vol. 2, arXiv: astro-ph/0310342
332. V. Sahni, A. Starobinsky, Int. J. Mod. Phys. D **9**, 373 (2000)
333. I.L. Shapiro, J. Solà, Phys. Lett. B **475**, 236 (2000)
334. T.R. Mongan, Gen. Relativ. Gravit. **33**, 1415 (2001)
335. S. Weinberg, Phys. Rev. D **61**, 103505 (2000)
336. S. Weinberg, Rev. Mod. Phys. **61**, 1 (1989)
337. W. Fischler, I. Klebanov, J. Polchinski, L. Susskind, Nucl. Phys. B **237**, 157 (1989)
338. S. Coleman, Nucl. Phys. B **310**, 643 (1988)
339. S. Coleman, Nucl. Phys. B **307**, 867 (1988)
340. S. Weinberg, Phys. Rev. Lett. **59**, 2607 (1987)
341. E. Baum, Phys. Lett. B **133**, 185 (1984)
342. S.W. Hawking, in *Shelter Island II – Proceedings of the 1983 Shelter Island Conference on Quantum Field Theory and the Fundamental Problems of Physics*, ed. by R. Jackiw et al. (MIT Press, Cambridge, 1995)
343. S.W. Hawking, Phys. Lett. B **134**, 403 (1984)
344. Ya.B. Zeldovich, Sov. Phys., Usp. **11**, 382 (1968) [Usp. Fiz. Nauk **95**, 209 (1968)]
345. C.P. Burgess et al., Strings, branes and cosmology: What can we hope to learn? arXiv: hep-th/0606020
346. C.P. Burgess et al., Supersymmetric large extra dimensions and the cosmological constant problem, arXiv: hep-th/0510123
347. T. Padmanabhan, Int. J. Mod. Phys. D **15**, 1659 (2006)
348. L. Parker, A. Raval, Phys. Rev. D **62**, 083503 (2000)
349. L. Parker, A. Raval, Phys. Rev. Lett. **86**, 749 (2001)
350. V. Blanloeil, B.F. Roukema (eds.), *Cosmological Topology in Paris 1998*, arXiv: astro-ph/0010170. See also Class. Quantum Gravity **15** (1998)

351. E. Elizalde, Phys. Lett. B **516**, 143 (2001)
352. L. Parker, A. Raval, Phys. Rev. D **62**, 083503 (2000)
353. L. Parker, A. Raval, Phys. Rev. Lett. **86**, 749 (2001)
354. E. Elizalde, J. Phys. A **34**, 3025 (2001)
355. E. Elizalde, J. Math. Phys. **35**, 3308 (1994)
356. E. Elizalde, J. Math. Phys. **35**, 6100 (1994)
357. E. Elizalde, Int. J. Mod. Phys. A **25**, 2345 (2010)
358. E. Elizalde, EAS Publ. Ser. **30**, 149 (2008)
359. E. Elizalde, J. Phys. Conf. Ser. **161**, 012019 (2009)
360. E. Elizalde, AIP Conf. Proc. **1115**, 123 (2009)
361. E. Elizalde, J. Phys. A **41**, 164061 (2008)
362. E. Elizalde, J. Phys. A **40**, 6647 (2007)
363. E. Elizalde, AIP Conf. Proc. **905**, 50 (2007)
364. E. Elizalde, PoS IC **2006**, 008 (2006)
365. E. Elizalde, AIP Conf. Proc. **878**, 232 (2006)
366. E. Elizalde, S. Nojiri, S.D. Odintsov, S. Ogushi, Phys. Rev. D **67**, 063515 (2003)
367. E. Elizalde, S. Nojiri, S.D. Odintsov, Phys. Rev. D **70**, 043539 (2004)
368. N. Boulanger, T. Damour, L. Gualtieri, M. Henneaux, Nucl. Phys. B **597**, 127 (2001)
369. A. Sugamoto, Gravit. Cosmol. **9**, 91 (2003)
370. N. Arkani-Hamed, A.G. Cohen, H. Georgi, Phys. Rev. Lett. **86**, 4757 (2001)
371. N. Arkani-Hamed, H. Georgi, M.D. Schwartz, Ann. Phys. (N.Y.) **305**, 96 (2003)
372. G. Cognola, E. Elizalde, S. Zerbini, Phys. Lett. B **624**, 70 (2005)
373. G. Cognola, E. Elizalde, S. Nojiri, S.D. Odintsov, S. Zerbini, Mod. Phys. Lett. A **19**, 1435 (2004)
374. N. Kan, K. Shiraishi, Class. Quantum Gravit. **20**, 4965 (2003)

Index

A

Abscissa of convergence, 8, 56
Accelerated expansion, 205
Additional contribution, 33, 34, 49, 151
Analytic continuation, 8, 12–14, 23, 29, 40, 50, 61, 63, 68, 69, 71, 76, 84, 95, 106, 108, 109, 112, 115, 119, 135, 147, 150–152, 184, 192, 193, 198, 199
Argument principle, 154–156, 163
Asymptotic expansion, 23–25, 27, 29, 30, 48, 51, 52, 59, 84–86, 89, 119, 123, 125

B

Background, 6, 24, 29, 101, 148, 175, 176, 178, 179, 191, 194, 200
Beta function, 109
Binomial expansion, 86, 120, 128, 138, 140, 194
Bose gas, 31
Boundary conditions, 6, 7, 106, 107, 110, 112, 114, 115, 118, 179–181, 186, 191
Braneworld, 203
Braneworlds, 210

C

Casimir effect, 12, 29, 66, 95, 96, 100–102, 105–107, 126
Casimir energy, 8, 102–104, 109, 118–120, 147, 152–154, 157–160, 162, 163, 178, 189
Casimir energy and cosmological constant, 207
Cauchy formula, 155
Chowla–Selberg formula, 67, 72, 76, 141
Color medium, 153
Computer, 157, 162
Contour at infinity, 42, 50, 51, 57
Contour integral, 32, 70, 71, 151, 153–155

Cosmological constant, 203
Covariant regularization, 148, 150
Critical behavior, 119
Critical point, 133
Critical radius, 185
Curvature of the space, 204

D

Deconstruction, 213
Degeneracy, 54, 123, 129, 154–156, 163
Determinant, 29, 120, 134–137, 140, 151, 152, 191
Dielectric, 98, 99, 160, 218
Differential operator, 6, 7, 29, 59, 66, 134, 135, 151, 152
Dimensional regularization, 11, 12
Dirac, 148, 182
Dirichlet boundary conditions, 110, 122
Dirichlet series, 8, 83
Discrete sum, 30
Dispersion function, 154, 156, 157, 163
Dixmier trace, 18
Domain, 7, 8, 51, 53, 76, 134
dS and AdS worldbranes, 203

E

Effective action, 60, 65, 105, 175–177
Eigenfrequencies, 30, 103, 154–156, 163
Eisenstein series, 138, 140, 219
Electromagnetic field, 97, 98, 101, 102, 106, 107, 114, 118
Elliptic operator, 49, 51
Energy-momentum tensor, 100, 105
Epstein zeta function, 7, 14, 23, 29, 30, 67, 68, 100, 113, 123, 192
Epstein–Hurwitz function, 39, 49, 69, 72, 74, 83, 84

E. Elizalde, *Ten Physical Applications of Spectral Zeta Functions*,
Lecture Notes in Physics 855,
DOI 10.1007/978-3-642-29405-1, © Springer-Verlag Berlin Heidelberg 2012

Euler, 127
Euler–Mascheroni, 32
Explicit calculation of the zeta function, 20
Extended Chowla–Selberg formula, 76, 77
Extension of the original CS formula to higher
    dimensions, 78

F
Faddeev–Popov, 179
Finite temperature, 141, 154, 160, 162, 163,
    189
Fractional values, 84, 119
Fujikawa method, 147, 148
Functional equation, 20

G
Gamma function, 53, 70, 105
Gauge condition, 117
Gauge theories, 147, 148, 151
Generalized operator regularization, 21
Geometric frequency, 160
Gravity, 29, 119, 134, 135, 140, 175, 177, 178,
    219
Green function, 100, 102

H
Hadamard regularization, 164, 170
Hadamard regularization of the Casimir effect,
    171
Hadamard regularized integral, 170
Hadamard's finite part, 172
Hadron, 153
Hamiltonian, 6–8, 30, 95, 103, 137, 140
Hardy–Ramanujan, 55
Harmonic oscillator, 30
Hemisphere, 98, 120
High-momentum approach, 126
Hurwitz zeta function, 23, 24, 27, 30, 38, 66,
    83, 84, 89, 119, 129, 130
Hypertorus, 115

I
Infinite contribution, 168
Inhomogeneity, 138, 160, 193
Integer part, 121
Interchange of sums order, 31, 71

J
Jacobi theta function, 36, 56, 58, 69, 112, 193

K
Kaluza–Klein, 119, 125, 126
Kelvin function, 194
Klein–Gordon equation, 103, 106, 107

L
Lagrange multipliers, 60
Lagrangian, 148, 191
Laplacian, 55, 120, 122, 123
Light, 96, 125, 126, 147, 153, 190
Low-momentum approach, 127, 130

M
Maple, 190
Mass matrix, 148
Massless scalar field, 107, 110, 117, 120, 123,
    189, 190, 198
Mathematica, 28, 190
Mathematical boundary conditions, 165, 202
Matsubara frequencies, 160, 162
Mellin transform, 43, 49, 51, 52, 54–56, 59,
    66, 70, 151, 192, 194, 216
Membrane, 29, 30, 49, 50, 60, 66, 99, 175
Meromorphic function, 8, 30, 63, 76, 104, 151,
    155
Metric, 6, 60, 61, 98, 126, 134, 135, 159, 176,
    199
Missing term, 36
Modified Bessel function, 40, 46, 73, 75, 109,
    194
Multi-graviton models, 211
Multi-index, 7
Multi-series, 7, 29, 30, 41
Multiplicative anomaly, 19
Multiplicities, 154, 163
Mystery of the Casimir effect, 100, 101

N
Nambu–Goto, 14, 60, 180
Neumann boundary conditions, 107, 114, 120
Non-polynomial term, 49
Non-zero temperature, 14, 102, 131
Noncommutative torus, 141
Normal ordering, 203
Number theory, 29, 67, 72, 83, 84, 119
Numerical computation, 30, 72, 86, 92, 123
Numerical estimates, 57
Numerical integration, 59

O
Operator expansion, 148
Operator regularization, 21
Optimal truncation, 50, 85, 127

P
p-brane, 30, 49, 60, 175, 178–180, 182
Parallel plates, 99, 100, 117, 118, 156
Partition function, 49, 50, 60
Pauli–Villars, 147, 148, 150, 151

Periodic boundary conditions, 108
Phase transition, 133
Piecewise uniform, 147, 152, 153, 162
Pochhammer's symbol, 25
Poisson formula, 55
Poisson summation formula, 56, 162
Poles, 34, 35, 47, 53, 57, 63, 84, 105, 119, 120, 122, 125, 155, 156, 192
Principal part prescription, 84, 104, 119, 124

Q
Quantum fluctuations, 179, 191, 194, 198, 200
Quantum system, 6

R
Radiation gauge, 117
Reality of zero point fluctuations, 202
Rectangle, 122, 123
Reflection formula, 20, 36
Refractive index, 153
Regularization theorem, 23, 29, 30, 41, 42, 49–51, 65, 66, 111, 112, 151
Regularized current, 150
Regulator, 104, 147, 148, 150, 151
Remainder, 50, 51, 54, 55, 57
Riemann generalized, 7
Riemann sphere, 120, 123, 124
Riemann zeta function, 8, 104, 108
Rigid membrane, 175, 178, 185

S
Saddle point, 179, 182, 183
Scalar field, 101, 103, 106, 107
Simple and double poles of the zeta function, 144
Sound, 13, 147, 153, 155
Spectral, 54, 135
Spectrum, 7, 8, 50, 65, 66, 95, 102, 110, 122, 125, 155, 181, 182
Spherical compactification, 30, 119, 125–127
Spontaneous compactification, 29, 175
Square, 73, 123, 190, 198
Stability, 29, 96, 175, 178

Stability analysis, 29
Stationary, 176–178
String, 14, 29, 30, 49, 50, 60, 66, 147, 152–157, 160–163, 175, 178–180, 182, 183, 185, 186
String physics, 179
Supergravitons, 203
Supermembrane, 50
Superstring, 50
Surface term, 154, 159
Symmetry breaking, 13, 125, 189, 190, 198, 200
Symmetry restoration, 189, 199, 200

T
Tachyon, 178, 179, 182, 185
Thermal frequency, 160
Topological mass, 189, 194, 195
Topology of space, 204
Toroidal compactification, 30
Torus, 83, 108, 189, 190, 196, 199, 200
Truncated, 121, 130

V
Vacuum energy, 66, 100–106, 108, 110, 114–116, 118, 120, 123, 124
van der Waals force, 96–98, 101, 102

W
Wheeler–De Witt, 119, 134, 137, 140
Wodzicki residue, 18

Z
Zero point energy, 156
Zero temperature, 67, 119, 153, 154
Zeta function, 7
Zeta function of a ΨDO, 17
Zeta function regularization, 6, 12–14, 24, 29, 30, 34, 41, 49, 65, 66, 120, 126, 153, 175
Zeta function renormalization, 12
Zeta regularized determinant, 18